# Industrial and Applied Mathematics

**Industrial and Applied Mathematics,** a Scopus-indexed book series, publishes high-quality research-level monographs, lecture notes and contributed volumes focusing on areas where mathematics is used in a fundamental way, such as industrial mathematics, bio-mathematics, financial mathematics, applied statistics, operations research and computer science.

S. A. Mohiuddine · Anupam Das ·
Dragan S. Djordjević
Editors

# Positive Operators and Fixed-Point Theorems with Applications

Springer

*Editors*
S. A. Mohiuddine
Department of General Required Courses
Mathematics, The Applied College
King Abdulaziz University
Jeddah, Saudi Arabia

Operator Theory and Applications
Research Group, Department
of Mathematics, Faculty of Science
King Abdulaziz University
Jeddah, Saudi Arabia

Dragan S. Djordjević
Faculty of Sciences and Mathematics
Department of Mathematics
University of Niš
Niš, Serbia

Anupam Das
Department of Mathematics
Cotton University
Guwahati, Assam, India

ISSN 2364-6837 ISSN 2364-6845 (electronic)
Industrial and Applied Mathematics
ISBN 978-981-96-2475-1 ISBN 978-981-96-2476-8 (eBook)
https://doi.org/10.1007/978-981-96-2476-8

Mathematics Subject Classification: 47B38, 47H10, 41A36, 47H08, 54H25, 58C30, 58J20, 41A10, 41A25, 26A33, 45G15, 34L30



This Springer imprint is published by the registered company Springer Nature Singapore Pte Ltd.
The registered company address is: 152 Beach Road, #21-01/04 Gateway East, Singapore 189721, Singapore

# Preface

This book features original chapters on positive operators, fixed-point theorems and illustrates their involvement in various applications. This book introduces some positive operators and studies their approximation properties such as Korovkin-type theorems, Voronovskaja-type results, rate of convergence and other related results as well as covers the study of existence of solutions of some differential and integral equations in Banach spaces in conjunction with Darbo-type fixed point theorems. Each chapter describes the problem of current importance, and to summarize ways of their solution and possible applications. The presentation of the chapters is clear and self-contained throughout the entire book. The list of chapters is arranged alphabetically by the last names of the first author of each chapter.

Chapter Convergence of a Class of Szász–Kantorovich Operators Based on 2D Appell Polynomials is devoted to study some direct theorems and local approximation results of a family of Kantorovich operators including 2D Appell polynomials. Next, a connection between summability methods, the statistical convergence of the related operators is established. Moreover, certain graphical examples to show the accuracy and efficiency of the proposed operators in terms of selected parameters are investigated.

Chapter Convergence Results for Nonlinear Generalized Bernstein Type Operators is devoted to construct a new sequence of generalized non-linear Bernstein-type operators. For these operators, a Kelisky–Rivlin type theorem is proved and their rate of convergence is determined. Also, a fixed point theorem and a generalized JS-contraction mapping are introduced.

Chapter Existence of Solutions for Infinite System of Nonlinear $q$-Fractional Boundary Value Problem via Generalized Darbo's Fixed Point Theorem uses a contraction operator to prove a generalized Darbo-type fixed point theorem with the intention to study the existence of solutions of infinite system of nonlinear $q$-fractional boundary value problem in Banach spaces formulated with the help of measure of noncompactness. Moreover, a suitable example is presented in order to illustrate the existence result.

Chapter On the Solutions of a Nonlinear Fractional Integral Equation Using the Darbo Fixed Point Theorem generalizes a fixed point theorem using a newly designed contraction operator. In addition, this chapter describes the solvability of fractional integrals based on this generalized fixed point theorem along with appropriate examples.

Chapter The Existence Solutions of an Integral Equation Using the Coupled Measure of Noncompactness is devoted to study the results of fixed point theorem in measure of noncompactness and the coupled measure of noncompactness. Some results on the measure of noncompactness and the coupled measure of noncompactness for the family of contraction operators are proved. Moreover, an application with an example is provided to illustrate the results.

In Chap. Generalized Bernstein Operators Associated with Multiple Shape Parameters, a novel extension of Bernstein-type operators are introduced based on multiple shape parameters $\lambda_i, i = 1, 2, \ldots n$. Convergence properties of the operators are investigated and local and global approximation theorems are established via Ditzian–Totik modulus of smoothness of first and second order. The rate of convergence is estimated with the help of Lipschitz-type function involving two parameters. Furthermore, statistical approximation properties are given using weighted mean matrix method and Voronovskaja-type asymptotic approximation results are presented. Finally, some illustrative graphs and numerical examples are provided to show the efficiency, applicability and validity of proposed operators.

In Chap. Approximation by $q$-analogue of Modified Szász Operators Generated by Dunkl Exponential Functions, a modification of Szász operators generated by Dunkl exponential functions via $q$-calculus based on a continuously differentiable function $T$ on $[0, \infty)$ is defined. The uniform approximation and weighted approximation of aforesaid operators are obtained. Moreover, the rate of convergence of our operators for functions belonging to the Lipschitz class including some other related approximation results is presented.

Chapter A Study on Range and Kernel of Toeplitz and Hankel Operators on the Bergman Space establishes the inclusion relationship between the kernel and range of little Hankel operators and Toeplitz operators on the Bergman space. Further, the conditions for finite rank operators and invertibility of operators on $A^2(\mathbb{D})$ are expounded. Moreover, some necessary and sufficient conditions are explored under which the bounded linear operators on the Bergman space $A^2(\mathbb{D})$ are positive.

In Chap. Fundamental Theorem of Approximation for Linear and Non-linear Sampling Kantorovich Operators, a brief introduction of sampling theory and how it connects with the theory of sampling Kantorovich operators are given. The main task of this chapter is to prove the fundamental theorem of approximation for different operators related to sampling Kantorovich operators. More preciously, the approximation theorem is proved for linear and non-linear using sampling Kantorovich theory. Eventually, an example of the justification of the fundamental theorem is illustrated.

Chapter Fixed-Point Method: Stability Results of Functional Equation Deriving from Additive-Quadratic Mappings in 2-Banach Spaces is devoted to investigate the Ulam stability of the quadratic-additive functional equation in 2-Banach spaces using direct method and fixed point method. The results obtained have a practical applications, demonstrating that the stability of the quadratic-additive functional equation can be controlled through the sums and products of powers of norms.

We wish to express our gratitude to the authors who have contributed to this book. We would like to thank our family for moral support during the preparation of this book. Finally, we are also very thankful to Mr. Shamim Ahmad, Executive Editor of Mathematics in Springer, for taking interest in publishing this book.

Jeddah, Saudi Arabia S. A. Mohiuddine
Guwahati, India Anupam Das
Niš, Serbia Dragan S. Djordjević

# Contents

# About the Editors

**S. A. Mohiuddine** is a full professor of Mathematics at King Abdulaziz University, Jeddah, Saudi Arabia. An active researcher, he has co-authored five books: *Convergence Methods for Double Sequences and Applications* (Springer, 2014), *Advances in Summability and Approximation Theory* (Springer, 2018), *Approximation Theory, Sequence Spaces and Applications* (Springer, 2022), *Soft Computing Techniques in Engineering, Health, Mathematical and Social Sciences, Sequence Space Theory with Applications*, and a number of chapters in several edited book and has contributed more than 150 research papers to various leading journals. He is the referee of many scientific journals and member of the editorial board of various scientific journals, international scientific bodies and organizing committees. He has visited several international universities including Imperial College London, UK. He was a guest editor of a number of special issues in some journals. His research interests are in the fields of sequence spaces, statistical convergence, matrix transformation, measures of non-compactness and approximation theory. His name has been appeared in the list of World's Top 2% Scientists (for the 3 consecutive years 2020–2022) prepared by Stanford University, California, through Scopus data provided by Elsevier.

**Anupam Das** is an Associate Professor of Mathematics at Department of Mathematics, Cotton University, Guwahati, Assam, India. Prior to this, he worked as Assistant Professor in the Department of Mathematics, Rajiv Gandhi University, Doimukh, Arunachal Pradesh, India. He received his Ph.D. in Mathematics from the Department of Mathematics, Rajiv Gandhi University, Doimukh, Arunachal Pradesh, India. His research interests include summability theory, fixed point theory, best proximity point theory and study of fractional integral and differential equations. He has published over 40 articles in various international journals. He has successfully guided a Ph.D. student in the area of fixed point theory. Dr. Das has qualified several national-level examinations in mathematics in India.

**Dragan S. Djordjević** was born on March 17, 1970. He received his Bachelor (1992), Master (1996) and Ph.D. (1998) degree in mathematics at the University of Niš, Serbia. Dragan Djordjević work at the University of Niš continuously since 1992, and from 2006 as a full professor at the Faculty of Sciences and Mathematics. He taught several courses in various levels of education (Functional analysis, Operator theory, Complex analysis, Measure and integration, Algebras of operators). He was dean of the Faculty of Sciences and Mathematics, and a member of several scientific boards of the University of Niš (Senat, Scientific Council for Sciences and Mathematics), Ministry of Education, Science and Technological Development of Serbia (Scientific Council for Mathematics, Mechanics and Computer Science, Head of the Council), as well as in the Republic of Serbia (National Council for Higher Education). He was a head and a member of several scientific projects financed by the Ministry of Education, Science and Technological Development of Serbia. He is a member and vice president of the association Serbian Scientific Mathematical Society. Dragan Djordjević supervised 10 Ph.D. thesis (8 in Niš, 1 in Belgrade and 1 in Skopje). He published 130 scientific papers, including 104 papers in JCR journals. He collaborated with more that 30 authors, most of them were his students. His results are cited approximately 1400 times, and the Hirsch index is h = 22. Dragan Djordjević served as the invited speaker at several mathematical conferences (Korea, Japan, Turkey), He also organized conferences in mathematical or functional analysis in Niš and Novi Sad. He also reviewed papers for many mathematical journals, including Mathematical Reviews. His scientific work includes operator theory, with some aspects of functional analysis, linear algebra, general algebra and numerical analysis. His main focus is on operators on Banach and Hilbert spaces, their ordinary and generalized invertibility properties, Fredholm and spectral theory, as well as applications to general ring theory and operator equations in various structures. He also has some results related to majorization and doubly stochastic operators, theory of spaces with indefinite inner products and Hilbert C*-modules. Dragan Djordjević is a member of boards of some mathematical journals: Filomat (one of editors in chief), Functional Analysis, Approximation and Computation (founder and editor in chief), Applied Mathematics and Computer Science (founder and one of editors in chief), Advances in the Theory of Nonlinear Analysis and its Application (member of the board), Publications de l'Institute Mathematique (member of the board); Advances in Operator Theory (member of the board).

# Convergence of a Class of Szász–Kantorovich Operators Based on 2D Appell Polynomials

**Mahvish Ali, Reşat Aslan, and Faruk Özger**

**Abstract** In this chapter, we aim to study some direct theorems and local approximation results of a family of Kantorovich operators including 2D Appell polynomials. Next, we establish a connection between summability methods, and the statistical convergence of the related operators. Moreover, we investigate certain graphical examples to show the accuracy and efficiency of the proposed operators in terms of selected parameters.

**Keywords** 2D Appell polynomials · Kantorovich operators · Peetre's K-functional · Weighted statistical convergence

**AMS Subject Classification (2020)** 41A10 · 41A25 · 41A36

## 1 Introduction

The field of approximation theory, which is a sub-branch of functional analysis and still has many utilizations in disciplines such as physics and engineering. Especially in the early nineteenth century, the problem of approaching a certain quantity has been one of the oldest problems faced by mathematicians. One of the first pioneering works of approximation theory was put forward in 1854 by the famous Russian mathematician P. L. Chebyshev: It can be expressed as "*The treatment of the problem of approximation to continuous functions with algebraic polynomials*". In 1885, German mathematician Weierstrass [1] proposed and proved the following well-known approximation theorem as follows.

M. Ali
Mathematics Section, Women's College, Aligarh Muslim University, Aligarh, India

R. Aslan (✉)
Department of Mathematics, Faculty of Sciences, Van Yuzuncu Yil University, Van, Türkiye
e-mail: resat63@hotmail.com

F. Özger
Department of Computer Engineering, Iğdır University, Iğdır, Türkiye

 
S. A. Mohiuddine et al. (eds.), *Positive Operators and Fixed-Point Theorems with Applications*, Industrial and Applied Mathematics,
https://doi.org/10.1007/978-981-96-2476-8_1

**Theorem 1.1** *Assume that a function* $g$ *is the continuous function on* $[u, v]$. *For given any* $\varepsilon > 0$, *there is polynomial* $P_m$ *sufficiently of high degree* $m$ *such that*

$$|g(z) - P_m(z)| \leq \varepsilon \text{ for } u \leq z \leq v, u, v \in \mathbb{R}.$$

L. Fejer, C. Runge, H. L. Lebesgue and S. N. Bernstein, some of the famous mathematicians of that period, found the proof of above-mentioned theorem too long and complex, so the studies on the proof of this theorem gained momentum and many alternative proofs were obtained. In conclusion, the most elegant and magnificent of them were presented by Bernstein [2] via polynomial sequence in 1912 as

$$B_n(f; z) = \sum_{j=0}^{n} q_{n,j}(z) f(\frac{j}{n}),\ z \in [0, 1],\ f \in C[0, 1], \tag{1.1}$$

where $n \in \mathbb{N}$ and $q_{n,j}(z) = \binom{n}{j} z^j (1-z)^{n-j}$.

In 1930, an integral modification of the Bernstein polynomials for the space of all Lebesgue integrable functions on [0, 1] was considered by Kantorovich [3] as follows:

$$K_n(f; z) = (n+1) \sum_{j=0}^{n} q_{n,j}(z) \int_{\frac{j}{n+1}}^{\frac{j+1}{n+1}} f(t)dt,\ z \in [0, 1]. \tag{1.2}$$

In [4], O. Szász offered and studied the following linear positive operators, for any $n \in \mathbb{N}$ and for the bounded functions

$$S_n(f; z) = \sum_{j=0}^{\infty} s_{n,j}(z) f\left(\frac{j}{n}\right), z \geq 0, \tag{1.3}$$

where $f(z) \in C[0, \infty)$ and $s_{n,j}(z) = e^{-nz} \frac{(nz)^j}{j!}$.

In 1953, Korovkin [5] simplified Bernstein's terrific proof of these sequences of polynomials and based on this they created the concept of "*Linear positive operators and their approximation*".

In addition, Appell polynomials [6], which have a separate accumulation in the literature, are a class of polynomials that is useful in many branches of mathematics, and contain algebra, analysis and probability theory. The set of all Appell sequences is closed under the operation of umbral compositions of polynomial sequences. The Appell polynomial form is an abelian group under the operation of umbral structure of polynomial sequences. The polynomial sequence $(u_m(x))_{m=0}^{\infty}$ means a sequence in which the degree of $u_m(x)$ is $m$. The umbral composition $((u_m \circ v)(x))_{m=0}^{\infty}$ of two polynomial sequences $(u_m(x))_{m=0}^{\infty}$ and $(v_m(x))_{m=0}^{\infty}$, where for every $m$ we get $u_m(x) = \sum_{\zeta=0}^{m} u_{m\zeta} x^{\zeta}$, which is given by $(u_m \circ v)(x) = \sum_{\zeta=0}^{m} u_{m\zeta} v_{\zeta}(x)$. An Appell sequence is a polynomial sequence $(u_m(x))_{m=0}^{\infty}$ for which $u_m^{'}(x) = mu_{m-1}(x)$ for $m \geq 1$. A sequence of binomial type is a polynomial sequence $(u_m(x))_{m=0}^{\infty}$ for which $u_m(x+y) = \sum_{\zeta=0}^{m} \binom{m}{\zeta} u_\zeta(x) u_{m-\zeta}(y)$ for $m \geq 0$. Umbral calculus is a theory of polynomials that count combinatorial objects. Properties of the Appell sequences are naturally handled within the framework of modern classical umbral calculus by Roman [7]. Appell polynomials are thoroughly explored by multiple perspectives due to their implementations in a diversity of fields. Finding recurrence correlations and differential equations by Appell polynomials is one component of this research. The popularity of Appell polynomials and their implementations in several areas have skyrocketed. The generalized Appell polynomials as approximation of 3D-mappings were improved by [8]. By employing methods of representation theory, they are also tools for applications in quantum physics [9].

The Appell polynomials can be given by one of two equivalent terms:

(i) $\frac{d}{dz} A_n(z) = n\ A_{n-1}(z), \qquad n \in \mathbb{N}_0$, or
(ii) for an exponential generating function with the following type:

$$A(v)\exp(zv) = \sum_{n=0}^{\infty} A_n(z) \frac{v^n}{n!},$$

where $A(v) = \sum_{n=0}^{\infty} A_n \frac{v^n}{n!}, \qquad A_0 \neq 0.$

The 2D Appell polynomials $G_k^{(r)}(z; a)$ for any $r \geq 2$ are defined with the following generating function [10]:

$$A(v)e^{zv+av^r} = \sum_{k=0}^{\infty} G_k^{(r)}(z; a) \frac{v^k}{k!}, \quad z, a \geq 0;\ t \neq 0 \tag{1.4}$$

where an analytic function $A(v)$ has the following power series:

$$A(v) = \sum_{m=0}^{\infty} \frac{\Omega_m}{m!} v^m, \quad A(0) \neq 0. \tag{1.5}$$

In [11], Jakimovski and Leviatan investigated a new form of linear positive operators on an unbounded interval via Appell polynomials as

$$L_n(f; z) = \frac{1}{e^{nz}\gamma(1)} \sum_{j=0}^{\infty} p_j(nz) f\left(\frac{j}{n}\right), \tag{1.6}$$

where $\gamma(t)e^{nt} = \sum_{j=0}^{\infty} p_j(t)n^j$ is generating function with Appell polynomials $p_j(t) \geq 0$ and $\gamma(v) = \sum_{j=0}^{\infty} a_j v^j$, $|v| < U$, $U > 1$ and $\gamma(1) \neq 0$. Note that, in case $\gamma(1) = 1$ in (1.6), we get $p_j(nz) = \frac{(nz)^j}{j!}$ and it reduces to the Szász [4] operators.

In the literature, the local approximation, weighted statistical approximation, Voronovskaja type asymptotic approximation and shape preserving properties of many generalizations and modifications of the above-mentioned polynomials have been studied in detail. Recently, Acar et al. [12] presented and worked the theoretical aspects of operators (1.6) and the convergence degree of quantitative estimates for the proposed operators. In [13], the authors considered an asymptotic relation of the derivatives for operators (1.6). Aslan [14–17] proposed Kantorovich versions of a new class Bernstein and Szász operators in terms of shape parameter $\lambda$. Mohiuddine et al. [18, 19] invented and attained some approximation properties of the Stancu variants of Bernstein–Kantorovich and Baskakov–Kantorovich operators on $[0, \infty)$ in terms of shape parameter $\alpha$. In [20], Büyükyazici et al. explored a generalization of the operators (1.6) and obtained some direct theorems and local approximation results. Ali and Kadak [21] studied Chlodowsky–Szász type of 2D Appell polynomials. A Kantorovich type generalization of Chlodowsky [22] type of operators (1.6) were established by Ari and Serenbay [23]. In [24, 25], Mursaleen et al. constructed and studied a new analogue of Stancu type Bernstein operators via $(p, q)$-integers and also Chlodowsky–Szász operators concerning Boas–Buck sort polynomials. Nasiruzzaman et al. [26, 27] established some parametric approximation outcomes of Phillips operators and obtained weighted and Voronovskaja type approximation for Szász–Kantorovich operators by $q$-Appell polynomials. Moreover, for some recent relevant papers we refer for the readers to [28–30].

According to this progress, we define the Szász–Kantorovich operators involving (1.4) as follows:

$$F_{n,G}(f; z) = \frac{n\, e^{-nz-a}}{A(1)} \sum_{k=0}^{\infty} \frac{G_k^{(r)}(nz, a)}{k!} \int_{\frac{k}{n}}^{\frac{k+1}{n}} f(s)ds, \tag{1.7}$$

where $a \geq 0$ and $G_k^{(r)}$ are the 2D Appell polynomials.

Outline of the chapter is as follows: In Sect. 2, we calculate several moment estimates and present the main Bohman–Korovkin theorem for operators (1.7). In Sect. 3, we provide local approximation properties, i.e. degree of convergence of the operators (1.7) for a class of Lipschitz-type functions and K-functional of Peetre. Next, in Sect. 4, we attain statistical approximation properties of the proposed operators and also the coincident degree of approximation using weighted $A$-statistical convergence of operators (1.7). Finally, we consider certain graphical examples and discuss the approximation behavior of operators (1.7) based on selected parameters.

## 2 Auxiliary Results

In this section, let us first to evaluate some moment estimates of operators (1.7). Let $e_u(s) = s^u$ be the test functions for $u = \{0, 1, 2, 3, 4\}$.

**Lemma 2.1** *The operators* $F_{n,G}(f; z)$ *constructed by* (1.7) *satisfy*

$$F_{n,G}(e_0; z) = 1, \tag{2.1}$$

$$F_{n,G}(e_1; z) = z + \frac{1}{n}\left(ar + \frac{1}{2} + \frac{A'(1)}{A(1)}\right). \tag{2.2}$$

$$\begin{aligned} F_{n,G}(e_2; z) &= z^2 + \frac{z}{n}\left(2 + 2ar + \frac{2A'(1)}{A(1)}\right) \\ &+ \frac{1}{n^2}\left(\frac{1}{3} + a^2r^2 + ar^2 + ar + \frac{2A'(1) + A''(1) + 2arA'(1)}{A(1)}\right), \end{aligned} \tag{2.3}$$

$$\begin{aligned} F_{n,G}(e_3; z) &= z^3 + \frac{z^2}{n}\left(\frac{9}{2} + 3ar + \frac{3A'(1)}{A(1)}\right) \\ &+ \frac{z}{n^2}\left(\frac{7}{2} + 9ar + 3a^2r^2 + 3ar(r-1) + \frac{9A'(1) + 6arA'(1) + 3A''(1)}{A(1)}\right) \\ &+ \frac{1}{n^3}\left(\nu_{a,r}^{(1)} + \frac{A'''(1) + 4A''(1) + 3A'(1) + 3arA''(1)}{A(1)}\right. \\ &\left.+ \frac{8arA'(1) + 3ar(r-1)A'(1) + 3a^2r^2A'(1)}{A(1)}\right), \end{aligned} \tag{2.4}$$

$$\begin{aligned} F_{n,G}(e_4; z) &= z^4 + \frac{z^3}{n}\left(8 + 4ar + \frac{4A'(1)}{A(1)}\right) + \frac{z^2}{n^2}\left(\nu_{a,r}^{(2)} + \frac{24A'(1) + 12arA'(1) + 6A''(1)}{A(1)}\right) \\ &+ \frac{z}{n^3}\left(\nu_{a,r}^{(3)} + \frac{4A'''(1) + 24A''(1) + 30A'(1) + 12arA''(1)}{A(1)}\right. \\ &\left.+ \frac{12a^2r^2A'(1) + 48arA'(1) + 12ar(r-1)A'(1)}{A(1)}\right) \\ &+ \frac{1}{n^4}\left(\nu_{a,r}^{(4)} + \frac{A^{(iv)}(1) + 8A'''(1) + 15A''(1) + 6A'(1)}{A(1)}\right. \\ &+ \frac{4arA'''(1) + 6a^2r^2A''(1) + 6ar(r-1)A''(1)}{A(1)} \\ &+ \frac{+24arA''(1) + 4a^3r^3A'(1) + 24ar(r-1)A'(1)}{A(1)} \\ &+ \frac{4ar(r-1)(r-2)A'(1) + 12a^2r^2(r-1)A'(1)}{A(1)} \\ &\left.+ \frac{24a^2r^2A'(1) + 24ar(r-1)A'(1) + 24arA'(1)}{A(1)}\right), \end{aligned} \tag{2.5}$$

*where*

$$\begin{aligned} \nu_{a,r}^{(1)} &= \frac{1}{4} + 3ar + 4ar(r-1) + 4a^2r^2 + 3a^2r^2(r-1) + ar(r-1)(r-2) + a^3r^3, \\ \nu_{a,r}^{(2)} &= 15 + 18ar + 6a^2r^2 + 6ar^2, \\ \nu_{a,r}^{(3)} &= 6 + 30ar + 12a^2r^2(r-1) + 24ar(r-1) + 4ar(r-1)(r-2) + 24a^2r^2 + 4a^3r^3, \end{aligned}$$

$$\begin{aligned}\nu_{a,r}^{(4)} =& \frac{1}{5} + 6ar + 15a^2r^2 + 13ar(r-1) + 8a^3r^3 + 2r(r-1) + 8ar(r-1)(r-2) \\ &+ 24a^2r^2(r-1) + 6a^3r^3(r-1) + ar(r-1)(r-2)(r-3) + 3a^2r^2(r-1)(2r-3) \\ &+ +a^2r^2(r-1)(r-2) + a^4r^4.\end{aligned}$$

Implementing the linearity property of the operators (1.7) and by Lemma 2.1, we get

$$\begin{aligned}F_{n,G}((e_1-z);z) &= \frac{1}{n}\left(ar + \frac{1}{2} + \frac{A'(1)}{A(1)}\right).\\ F_{n,G}((e_1-z)^2;z) &= \frac{z}{n} + \frac{1}{n^2}\left(\frac{1}{3} + a^2r^2 + ar^2 + ar + \frac{2A'(1)+A''(1)+2arA'(1)}{A(1)}\right).\\ F_{n,G}((e_1-z)^4;z) &= 3\frac{z^2}{n^2} + \frac{z}{n^3}\left(5 - 10ar + 18ar^2 - 16a^2r^2 + \frac{8A''(1)+18A'(1)+16arA'(1)}{A(1)}\right)\\ &+ \frac{1}{n^4}\left(\nu_{a,r}^{(4)} + \frac{A^{(iv)}(1)+8A'''(1)+15A''(1)+6A'(1)}{A(1)}\right.\\ &+ \frac{4arA'''(1)+6a^2r^2A''(1)+6ar(r-1)A''(1)}{A(1)}\\ &+ \frac{+24arA''(1)+4a^3r^3A'(1)+24ar(r-1)A'(1)}{A(1)}\\ &+ \frac{4ar(r-1)(r-2)A'(1)+12a^2r^2(r-1)A'(1)}{A(1)}\\ &\left.+ \frac{24a^2r^2A'(1)+24ar(r-1)A'(1)+24arA'(1)}{A(1)}\right).\end{aligned}$$

Now, we denote with $C_B[0,\infty)$ the set of all real-valued continuous and bounded functions $\kappa$ equipped with the following sup-norm:

$$\|\kappa\|_\infty = \sup_{z\in[0,\infty)} |\kappa(z)|.$$

In the following theorem, we will show that operators given by (1.7) are bounded and linear operators on $C_B[0,\infty)$.

**Theorem 2.1** *For $f \in C_B[0,\infty)$ and each $z \in [0,\infty)$, operators $F_{n,G}(.;z)$ satisfy*

$$\left|F_{n,G}(f;z)\right| \leq \|f\|_\infty.$$

***Proof*** Implementing the definition of (1.7) and Lemma 2.1 yields

$$\begin{aligned}\left|F_{n,G}(f;z)\right| &\leq \left|\frac{n\,e^{-nz-a}}{A(1)}\sum_{k=0}^{\infty}\frac{G_k^{(r)}(nz,a)}{k!}\int_{\frac{k}{n}}^{\frac{k+1}{n}} f(s)-f(z)ds\right|\\ &\leq \frac{n\,e^{-nz-a}}{A(1)}\sum_{k=0}^{\infty}\frac{G_k^{(r)}(nz,a)}{k!}\int_{\frac{k}{n}}^{\frac{k+1}{n}} |f(s)-f(z)|\,ds\end{aligned}$$

$$\leq \|f\|_\infty F_{n,G}(1; z)$$
$$\leq \|f\|_\infty ,$$

hence, we arrive at the proof. ■

**Theorem 2.2** *Let* $f \in C_B[0,\infty)$; *we get*

$$\lim_{n\to\infty} F_{n,G}(f; z) = f(z),$$

*which converges uniformly on* $[0,\infty)$.

***Proof*** Based on [5], it is adequate to verify that

$$\lim_{n\to\infty} \sup_{z\in[0,\infty)} \left|F_{n,G}(s^i; z) - z^i\right| = 0, \text{ for } i = 0, 1, 2.$$

Based on Lemma 2.1, the above relation follows indeed; therefore here we omitted the details. ■

## 3 Local Approximation Results

In this part, we derive the degree of approximation via ordinary moduli of continuity, a class of Lipschitz type functions and K-functional of Peetre. Now, let us present some notations and definitions.

The K-functional of Peetre is expressed by

$$K_2(\kappa,\zeta) = \inf_{\lambda\in C_B^2} \left\{\|\kappa-\lambda\|_\infty + \zeta \left\|\lambda''\right\|_\infty\right\}$$

where $\zeta > 0$ and $C_B^2 = \left\{\lambda \in C_B[0,\infty) : \lambda', \lambda'' \in C_B[0,\infty)\right\}$.

By [31], for an absolute constant $U > 0$, it follows

$$K_2(\kappa;\zeta) \leq U\omega_2(\kappa; \sqrt{\zeta}), \qquad \zeta > 0 \tag{3.1}$$

where the second-order moduli of smoothness for $\kappa \in C_B[0,\infty)$ is

$$\omega_2(\kappa;\zeta) = \sup_{0<a\leq\zeta} \sup_{z\in[0,\infty)} |\kappa(z+2a) - 2\kappa(z+a) + \kappa(z)| .$$

Further, with

$$\omega(\kappa;\zeta) := \sup_{0<a\leq\zeta} \sup_{z\in[0,\infty)} |\kappa(z+a) - \kappa(z)|$$

we represent the ordinary moduli of continuity of $\kappa \in C_B[0,\infty)$. If $\zeta > 0$, $\omega(\kappa;\zeta)$ satisfies the properties [32]. Additionally, we give the elements of Lipschitz type class with $Lip_\gamma(\eta)$, where $\gamma > 0$ and $0 < \eta \leq 1$. If the following inequality

$$|\kappa(t) - \kappa(z)| \leq \gamma\, |t - z|^{\eta}, \qquad (t, z \in \mathbb{R})$$

holds true, then a function $\kappa$ belongs to $Lip_\gamma(\eta)$.

**Theorem 3.1** *Let $f \in C[0,\infty)$. Then for all $z \in [0,\infty)$, operators $F_{n,G}(.;z)$ satisfy the following relation:*

$$\left|F_{n,G}(f;z) - f(z)\right| \leq 2\omega(f;\sqrt{\xi_{n,2}(z)})$$

*where $\xi_{n,2}(z) = F_{n,G}((e_1 - z)^2; z)$.*

***Proof*** From the definition of operators (1.7) and by applying the common property of modulus of continuity, one has

$$\left|F_{n,G}(f;z) - f(z)\right| \leq \frac{n\, e^{-nz-a}}{A(1)} \sum_{k=0}^{\infty} \frac{G_k^{(r)}(nz,a)}{k!} \int_{\frac{k}{n}}^{\frac{k+1}{n}} |f(s) - f(z)|\, ds$$

$$\leq \omega(f;\gamma)\left\{1 + \frac{1}{\gamma}\frac{n\, e^{-nz-a}}{A(1)} \sum_{k=0}^{\infty} \frac{G_k^{(r)}(nz,a)}{k!} \int_{\frac{k}{n}}^{\frac{k+1}{n}} |s - z|\, ds\right\}.$$

Utilizing the Cauchy–Schwarz inequality and Lemma 2.1 implies

$$\left|F_{n,G}(f;z) - f(z)\right| \leq \omega(f;\gamma)\left\{1 + \frac{1}{\gamma}\frac{n\, e^{-nz-a}}{A(1)} \sum_{k=0}^{\infty} \frac{G_k^{(r)}(nz,a)}{k!} \int_{\frac{k}{n}}^{\frac{k+1}{n}} (s - z)^2\, ds)^{\frac{1}{2}}\right\}$$

$$\leq \omega(f;\gamma)\left\{1 + \frac{1}{\gamma}\left(\xi_{n,2}(z)\right)^{\frac{1}{2}}\right\}.$$

Taking $\gamma = \left(\xi_{n,2}(z)\right)^{\frac{1}{2}}$ ends the proof. ■

**Theorem 3.2** *Let $f \in C[0,\infty)$, $\eta > 0$. Then for $\forall z \in [0,\infty)$ and $f \in Lip_\gamma(\eta)$, the following relation holds true:*

$$\left|F_{n,G}(f;z) - f(z)\right| \leq \gamma(\xi_{n,2}(z))^{\frac{\eta}{2}}.$$

***Proof*** In view of the linearity and positivity features of the operator $F_{n,G}(.; z)$, we get

$$\begin{aligned}\left|F_{n,G}(f;z)-f(z)\right| &\le F_{n,G}(\left|f(t)-f(z)\right|;z)\\ &= \frac{n\,e^{-nz-a}}{A(1)}\sum_{k=0}^{\infty}\frac{G_k^{(r)}(nz,a)}{k!}\int_0^1\left|f\left(\frac{k+t}{n}\right)-f(z)\right|dt\\ &\le \gamma\left\{\frac{n\,e^{-nz-a}}{A(1)}\sum_{k=0}^{\infty}\frac{G_k^{(r)}(nz,a)}{k!}\int_0^1\left|\frac{k+t}{n}-z\right|^{\eta}dt\right\}.\end{aligned}$$

In view of Hölder's inequality for $p_1 = \frac{2}{\eta}$ and $p_2 = \frac{2}{2-\eta}$ and Lemma 2.1, one has

$$\begin{aligned}&\left|F_{n,G}(f;z)-f(z)\right|\\ &\le \gamma\left\{\frac{n\,e^{-nz-a}}{A(1)}\sum_{k=0}^{\infty}\frac{G_k^{(r)}(nz,a)}{k!}\int_0^1\left(\frac{k+t}{n}-z\right)^2dt\right\}^{\frac{\eta}{2}}\left\{F_{n,G}(1^2;z)\right\}^{\frac{2-\eta}{2}}\\ &= \gamma\left\{F_{n,G}((e_1-z)^2;z)\right\}^{\frac{\eta}{2}}\left\{F_{n,G}(1^2;z)\right\}^{\frac{2-\eta}{2}}\\ &= \gamma(\xi_{n,2}(z))^{\frac{\eta}{2}}.\end{aligned}$$

Thus, one has the assertion. ■

**Theorem 3.3** *For each* $f \in C[0,\infty)$ *and* $z \in [0,\infty)$*, then*

$$\left|F_{n,G}(f;z)-f(z)\right| \le C\omega_2(f;\frac{1}{2}\sqrt{\xi_{n,2}(z)+(\xi_{n,1}(z))^2}+\omega(f;\left|\xi_{n,1}(z)\right|),$$

*where a constant* $C > 0$, $\xi_{n,1}(z) = F_{n,G}(e_1 - z; z)$ *and* $\xi_{n,2}(z)$ *defined by Theorem 3.1.*

***Proof*** Let us denote $\alpha_n(z) := z + \frac{1}{n}\left(ar + \frac{1}{2} + \frac{A'(1)}{A(1)}\right)$. For sufficiently large $n$, it is obvious that $\alpha_n(z) \in [0,\infty)$. Now, let us consider the following auxiliary operator:

$$\overset{\bullet}{F}_{n,G}(f;z) = f(z) + F_{n,G}(f;z) - f(\alpha_n(z)). \tag{3.2}$$

Lemma 2.1 yields

$$\overset{\bullet}{F}_{n,G}(l-z;z) = 0.$$

For any $\Gamma \in C^2[0,\infty)$, applying Taylor's formula it follows that

$$\Gamma(l) = \Gamma(z) + (l-z)\Gamma'(z) + \int_z^l (l-m)\Gamma''(m)dm. \tag{3.3}$$

Operating $\overset{\bullet}{F}_{n,G}(.;z)$ to (3.3), we arrive at

$$\overset{\bullet}{F}_{n,G}(\Gamma;z)-\Gamma(z)=\overset{\bullet}{F}_{n,G}((l-z)\Gamma'(z);z)+\overset{\bullet}{F}_{n,G}(\int_z^l(l-m)\Gamma''(m)dm;z)$$

$$=\Gamma'(z)\overset{\bullet}{F}_{n,G}(l-z;z)+F_{n,G}(\int_z^l(l-m)\Gamma''(m)dm;z)-\int_z^{\alpha_n(z)}(\alpha_n(z)-m)\Gamma''(m)dm$$

$$=F_{n,G}(\int_z^l(l-m)\Gamma''(m)dm;z)-\int_z^{\alpha_n(z)}(\alpha_n(z)-m)\Gamma''(m)dm.$$

Using Lemma 2.1, then

$$\left|\overset{\bullet}{F}_{n,G}(\Gamma;z)-\Gamma(z)\right|\le\left|F_{n,G}(\int_z^l(l-m)\Gamma''(m)dm;z)\right|+\left|\int_z^{\alpha_n(z)}(\alpha_n(z)-m)\Gamma''(m)dm\right|$$

$$\le F_{n,G}(\left|\int_z^l(l-m)\right|\left|\Gamma''(m)\right|dm;z)+\int_z^{\alpha_n(z)}|\alpha_n(z)-m|\left|\Gamma''(m)\right|dm$$

$$\le\left\|\Gamma''\right\|_\infty\left\{F_{n,G}((l-z)^2;z)+(\alpha_n(z)-z)^2\right\}$$

$$\le\left\|\Gamma''\right\|_\infty\left\{\xi_{n,2}(z)+(\xi_{n,1}(z))^2\right\}.$$

Taking Lemmas 2.1 and (3.2) into account, we get

$$\left|\overset{\bullet}{F}_{n,G}(f;z)\right|\le2\|f\|_\infty+\left|F_{n,G}(f;z)\right|$$

$$\le2\|f\|_\infty+\|f\|_\infty F_{n,G}(1;z)\le3\|f\|_\infty. \tag{3.4}$$

Also, by (3.2) and (3.4) it follows that

$$\left|F_{n,G}(f;z)-f(z)\right|\le\left|\overset{\bullet}{F}_{n,G}(f-\Gamma;z)-(f-\Gamma)(z)\right|$$

$$+\left|\overset{\bullet}{F}_{n,G}(\Gamma;z)-\Gamma(z)\right|+|f(z)-f(\alpha_n(z))|$$

$$\le4\|f-\Gamma\|_\infty+\left\{\xi_{n,2}(z)+(\xi_{n,1}(z))^2\right\}\left\|\Gamma''\right\|_\infty+\omega\left(f;\left|\xi_{n,1}(z)\right|\right).$$

Hereby, getting the infimum on the right-hand side over all $\Gamma\in C^2[0,\infty)$ and making use of (3.1) then

$$\left|F_{n,G}(f;z)-f(z)\right|\le4K_2(f;\frac{\left\{\xi_{n,2}(z)+(\xi_{n,1}(z))^2\right\}}{4})+\omega(f;\left|\xi_{n,1}(z)\right|)$$

$$\leq C\omega_2(f; \frac{1}{2}\sqrt{\xi_{n,2}(z) + (\xi_{n,1}(z))^2}) + \omega(f; |\xi_{n,1}(z)|),$$

and this ends the proof. ■

# 4 Approximation in Statistical Sense

Here, we study weighted statistical approximation properties of proposed operators and estimate the corresponding rate of convergence by weighted $A$-statistical convergence. We refer to [33–49] for the needed notions, notations and definitions.

Natural density of $K_n$ is denoted by $\delta(K) = \lim_n \frac{1}{n}|K_n|$ provided that limit exists, where $K_n = \{k \leq n : k \in K\}$, $K \subseteq \mathbb{N}_0 := \mathbb{N} \cup \{0\}$ and vertical bars denote cardinality of the enclosed set. A sequence $x = (x_n)$ of numbers is called statistically convergent to a number $L$, denoted by $st$-$\lim_n x = L$, if, for each $\epsilon > 0$,

$$\delta\{n : n \in \mathbb{N} \text{ and } |x_n - L| \geqq \epsilon\} = 0.$$

$A$-transform of $x$ denoted by $Ax := \{(Ax)_n\}$ is defined as $(Ax)_n = \sum_{k=0}^{\infty} a_{nk}x_k$ for a given non-negative infinite summability matrix $A = (a_{nk})$, $n, k \in \mathbb{N}$. It is provided series converges for every $n \in \mathbb{N}_0$. If $\lim_n (Ax)_n = L$ whenever $\lim_n x_n = L$, we say that $A$ is a regular method. Then sequence $x = (x_n)$ is said to be $A$-statistically convergent to $L$, denoted by $\mathrm{st}_A$-$\lim x = L$, provided that for each $\epsilon > 0$,

$$\sum_{k:|x_k - L| \geqq \epsilon} a_{nk} = 0 \qquad (n \to \infty).$$

$A$-statistical convergence becomes ordinary statistical convergence if we take $A = (C_1)$, the Cesaro matrix of order one, and it becomes classical convergence if we take $A = I$, the identity matrix. We know that every convergent sequence is statistically convergent to the same limit but not conversely.

Assume that $q = (q_n)$ is a sequence of non-negative numbers so that $q_0 > 0$ and $Q_n = \sum_{k=0}^{n} q_k \to \infty$ as $n \to \infty$. Then $x = (x_n)$ is called weighted $A$-statistically convergent to $L$, if, for every $\varepsilon > 0$,

$$\lim_{n\to\infty} \frac{1}{Q_n} \sum_{k=0}^{n} q_k \sum_{l:|x-L| \geqq \epsilon} a_{kl} = 0.$$

This relation is denoted by $S_A^{\widetilde{N}} - \lim x = L$ in this case. It is clear that weighted $A$-statistical convergence generalizes $A$-statistical convergence, which we recover by putting $q_n = 1$ for all $n \in \mathbb{N}$.

**Theorem 4.1** *Let $A = (a_{nk})$ be a weighted non-negative regular summability matrix for $n, k \in \mathbb{N}$ and $q = (q_n)$ be a sequence of non-negative numbers so that $q_0 > 0$*

*and* $Q_n = \sum_{k=0}^{n} q_k \to \infty$ *as* $n \to \infty$. *For any* $f \in C[0,1]$, *we have*

$$S_A^{\widetilde{N}} - \lim_{n\to\infty} \|F_{n,G}(f;z) - f(z)\|_{C[0,1]} = 0.$$

***Proof*** Consider the sequence of functions $e_j(z) = z^j$, where $j \in \{0,1,2\}$ and $z \in [0,1]$. It is sufficient to satisfy

$$S_A^{\widetilde{N}} - \lim_{n\to\infty} \|F_{n,G}(e_j;z) - e_j\|_{C[0,1]} = 0, \qquad j = 0,1,2.$$

The following relation is satisfied by Lemma 2.1:

$$S_A^{\widetilde{N}} - \lim_{n\to\infty} \|F_{n,G}(e_0;z) - e_0\|_{C[0,1]} = 0. \tag{4.1}$$

By Lemma 2.1

$$\begin{aligned}\|F_{n,G}(e_1;z) - e_1\|_{C[0,1]} &= \sup_{z\in[0,1]} \left| z + \frac{1}{n}\left(ar + \frac{1}{2} + \frac{A'(1)}{A(1)}\right) - z \right| \\ &\le \frac{2ar}{n} + \frac{1}{n} + \frac{2A'(1)}{nA(1)} = \Sigma_{a,r,n}\end{aligned}$$

is satisfied. Establishing the following sets

$$\begin{aligned}\Omega &:= \left\{n \in \mathbb{N} : \|F_{n,G}(e_1;z) - e_1\|_{C[0,1]} \geqq \bar{\epsilon}\right\}, \\ \Omega_1 &:= \left\{n \in \mathbb{N} : \Sigma_{a,r,n} \geqq \epsilon - \bar{\epsilon}\right\}\end{aligned}$$

and selecting a number $\epsilon > 0$ for a given $\bar{\epsilon} > 0$ such that $\epsilon < \bar{\epsilon}$, we see the inclusion $\Omega \subset \Omega_1$ is satisfied and

$$\frac{1}{Q_n}\sum_{k=0}^{n} q_k \sum_{l\in\Omega} a_{kl} \le \frac{1}{Q_n}\sum_{k=0}^{n} q_k \sum_{l\in\Omega_1} a_{kl} \qquad (\forall n). \tag{4.2}$$

The following relation is satisfied by taking limit as $n \to \infty$ in (4.2):

$$S_A^{\widetilde{N}} - \lim_{n\to\infty} \|F_{n,G}(e_1;z) - e_1\|_{C[0,1]} = 0. \tag{4.3}$$

We also have

$$\begin{aligned}\|F_{n,G}(e_2;z) - e_2\|_{C[0,1]} &= \sup_{z\in[0,1]} \left|F_{n,G}(e_2;z) - e_2\right| \\ &\le \frac{1}{n^2}\left(\frac{1}{3} + a^2r^2 + ar^2 + ar + \frac{2A'(1) + A''(1) + 2arA'(1)}{A(1)}\right) \\ &+ \frac{4}{n} + \frac{4ar}{n} + \frac{4arA'(1)}{A(1)} = \zeta_{a,r,n}.\end{aligned}$$

Since $S_A^{\widetilde{N}} - \lim_n \zeta_{a,r,n} = 0$, we get

$$S_A^{\widetilde{N}} - \lim_{n\to\infty} \|F_{n,G}(e_2; z) - e_2\|_{C[0,1]} = 0. \tag{4.4}$$

The desired result is attained via considering (4.1), (4.3) and (4.4) together. ■

An estimate for rate of weighted $A$-statistical convergence of $F_{n,G}(f; z)$ is given in the following theorem.

**Theorem 4.2** *Let $A = (a_{nk})$ be a weighted non-negative regular summability matrix. Assume the following condition:*

$w(f, \varphi_n) = [stat_A, q_n] - o\,(u_n)$ *on* $[0, 1]$, *where* $\varphi_n = \sqrt{\|F_{n,G}((s-z)^2; z)\|_{C[0,1]}}$.

*The following relation is satisfied for $f \in B[0, 1]$ :*

$$\|F_{n,G}(f; z) - f(z)\|_{C[0,1]} = [stat_A, q_n] - o\,(u_n).$$

***Proof*** If $f(x, y) \in C[0, 1]$, $\psi = \sup_{z\in[0,1]} |f(z)|$ and $z, t \in [0, 1]$, we have

$$\begin{aligned}
|F_{n,G}(f; z) - f(z)| &\leq |F_{n,G}\big(|f(t) - f(z)|; z\big) + \psi\ |F_{n,G}(1; z) - 1| \\
&\leq \omega(f, \delta)F_{n,G}\left(\frac{|t-x|}{\delta} + 1; z\right) \\
&= \omega(f, \delta)F_{n,G}(1; z) + \omega(f, \delta)\frac{1}{\delta^2}F_{n,G}\left((t-x)^2; z\right).
\end{aligned}$$

Taking $\delta := \varphi_n$ for each $n \in \mathbb{N}$, and supremum over $z \in [0, \infty)$ on both sides, we attain

$$\|F_{n,G}(f; z) - f(z)\|_{C[0,1]} \leq \omega(f, \varphi_n) + \omega(f, \varphi_n)\frac{1}{\varphi_n^2}\|F_{n,G}((t-x)^2; z)\|_{C[0,1]} = 2\omega(f, \varphi_n).$$

Constructing the following

$$\begin{aligned}
\mathcal{B} &= \left\{n : \|F_{n,G}(f; z) - f(z)\|_{C[0,1]} \geq \epsilon\right\}, \\
\mathcal{W} &= \left\{n : \omega(f, \varphi_n) \geq \frac{\epsilon}{2}\right\}
\end{aligned}$$

for a given $\epsilon > 0$, the following inequality is satisfied:

$$\frac{1}{u_n Q_n}\sum_{k=0}^{n}\sum_{l\in\mathcal{B}} q_k a_{kl} \leq \frac{1}{u_n Q_n}\sum_{k=0}^{n}\sum_{l\in\mathcal{W}} q_k a_{kl}.$$

The theorem is proved by

$$\|F_{n,G}(f;z) - f(z)\|_{C[0,1]} = [stat_A, q_n] - o\,(u_n). \qquad \blacksquare$$

## 5 Effect of Parameters on Approximation

Here, we consider the following function:

$$f(z) = \frac{z^2 + 1}{z + 2}$$

to see the effect of parameters on approximation results and discuss approximation behavior of $F_{n,G}$ for different values of $a$, $n$ and $\rho$. In Figs. 1, 2 and 3, we demonstrate approximation and error of approximation of $F_{n,G}$ changing the parameters $n$, $\rho$ and $a$, respectively. This concludes that the proposed operators approximate to a given function for different values of given parameters.

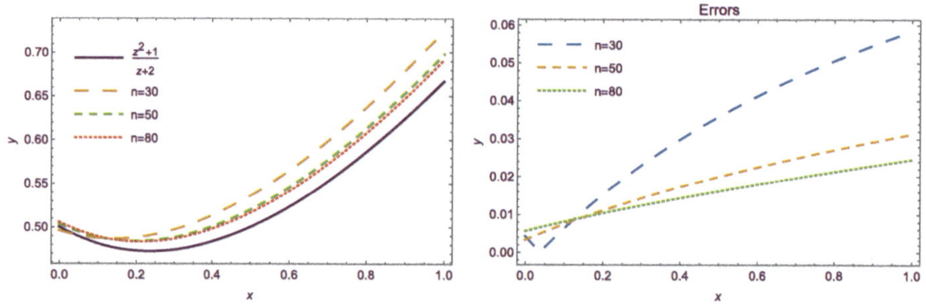


**Fig. 1** Approximation (left) and error of approximation (right) of $F_{n,G}$ for $a = 0.01$ and $\rho = 2$

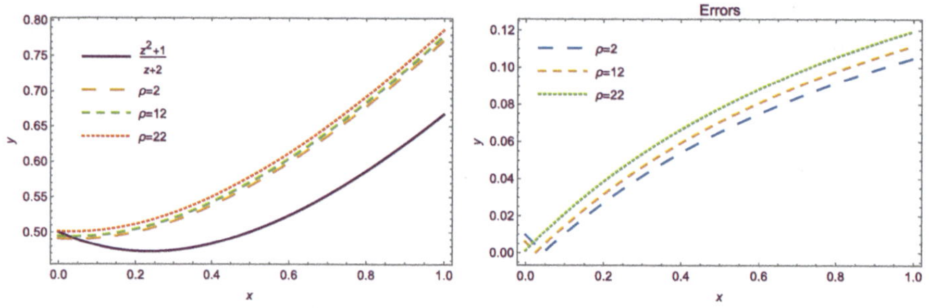


**Fig. 2** Approximation (left) and error of approximation (right) of $F_{n,G}$ for $n = 10$ and $a = 0.01$

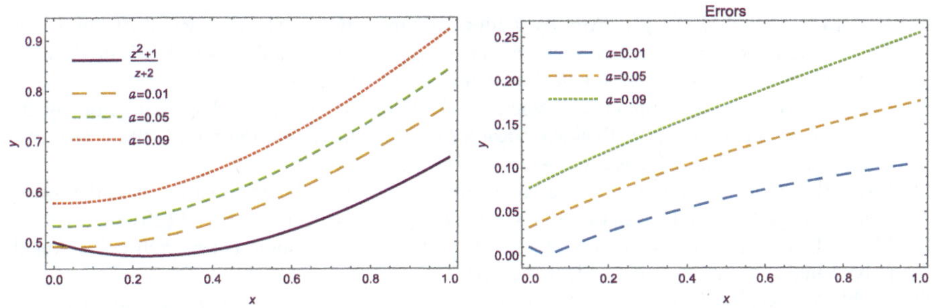


**Fig. 3** Approximation (left) and error of approximation (right) of $F_{n,G}$ for $n = 10$ and $\rho = 2$

# References


1. K. Weierstrass, Über die analytische darstellbarkeit sogenannter willkürlicher functionen einer reellen veränderlichen, in *Sitzungsberichte der Königlich Preußischen Akademie der Wissenschaften zu Berlin*, pp. 633–639 (1885)
2. S.N. Bernstein, Démonstration du théorème de Weierstrass fondée sur le calcul des probabilités. Comput. Commun. Soc. Mat. Charkow Ser. **13**, 1–2 (1912)
3. L.V. Kantorovich, Sur certains developements suivant les polynmes de la forme de S. Bernstein I, II. Dokl Akad Nauk SSSR *563*, 595–600 (1930)
4. O. Szász, Generalization of S. Bernstein's polynomials to the infinite intervals. J. Res. Nat. Bur. Stand. *45*, 239–245 (1950)
5. P.P. Korovkin, On convergence of linear positive operators in the space of continuous functions. Dokl Akad Nauk SSSR. **90**, 961–964 (1953)
6. P. Appell, Sur une classe de polynômes. Ann. Sci. École. Norm. Sup. **9**, 119–144 (1880)
7. S. Roman, *The Umbral Calculus* (Springer, New York, 2005)
8. H.R. Malonek, Falcão, M.I. 3D-mappings by means of monogenic functions and their approximation. Math. Methods Appl. Sci. **33**, 423–430 (2010)
9. S.T. Weinberg, *The Quantum Theory of Fields*, vol. 1 (Cambridge University Press, Cambridge, 1995)
10. G. Bretti, C. Cesarano, P.E. Ricci, Laguerre-type exponentials and generalized Appell polynomials. Comput. Math. Appl. **48**, 833–839 (2004)
11. A. Jakimovski, D. Leviatan, Generalized Szász operators for the approximation in the infinite interval. Mathematica (Cluj). **11**, 97–103 (1969)
12. E. Acar, A. İzgi, S. Serenbay Kirci, Note on Jakimovski-Leviatan operators preserving. Appl. Math. Nonlinear Sci. **4**, 543–550 (2019)
13. U. Abel, M. Ivan, Asimptotic expansion of the Jakimovski -Leviatan operators and their derivatives, in *Functions, Series, Operators*, Budapest, ed. by L. Leindler, F. Schipp, J. Szabados, pp. 103–119 (1999)
14. R. Aslan, Approximation properties of univariate and bivariate new class λ-Bernstein-Kantorovich operators and its associated GBS operators. Comput. Appl. Math. **42**, 34 (2023)
15. R. Aslan, Some approximation results on λ−Szasz-Mirakjan-Kantorovich operators. Fundam. J. Math. Appl. **4**, 150–158 (2021)
16. R. Aslan, Rate of approximation of blending type modified univariate and bivariate λ−Schurer-Kantorovich operators. Kuwait J. Sci. **51**(1) (2024), Article id:100168
17. R. Aslan, On a Stancu form Szász-Mirakjan-Kantorovich operators based on shape parameter λ. Adv. Stud.: Euro-Tbilisi Math. J. **15**(1), 151–166 (2022)

18. S.A. Mohiuddine, F. Özger, Approximation of functions by Stancu variant of Bernstein-Kantorovich operators based on shape parameter $\alpha$, Revista de la Real Academia de Ciencias Exactas. Físicas y Naturales. Ser. A. Matemáticas **114**, 70 (2020)
19. S.A. Mohiuddine, N. Ahmad, F. Özger, A. Alotaibi, B. Hazarika, Approximation by the parametric generalization of Baskakov-Kantorovich operators linking with Stancu operators. Iranian J. Sci. Technol. Trans.: Sci. **45**, 593–605 (2021)
20. I. Büyükyazici, H. Tanberkan, S. Serenbay Kirci, Ç. Atakut, Approximation by Chlodowsky type Jakimovski-Leviatan operators. J. Comput. Appl. Math. *259*, 153–163 (2014)
21. M. Ali, U. Kadak, Approximation by Szász-Chlodowsky type operators associating 2D-Appell polynomials. Z. Angew. Math. Mech. e202100229 (2022)
22. I. Chlodowsky, Sur le développement des fonctions définies dans un intervalle infini en séries de polynomes de MS Bernstein. Compos. Math. **4**, 380–393 (1937)
23. D.A. Ari, S. Serenbay Kirci, Approximation by Chlodowsky type Jakimovski-Leviatan operators. Filomat **33**, 2345–2353 (2019)
24. M. Mursaleen, K.J. Ansari, A. Khan, OME approximation results by (p, q)-analogue of Bernstein-Stancu operators. Appl. Math. Comput. **264**, 392–402 (2015)
25. N.L. Braha, V. Loku, M. Mursaleen, Chlodowsky-type Szász operators via Boas-Buck-type polynomials and some approximation properties. J. Inequal. Appl. **2023**, 1–23 (2023)
26. Md. Nasiruzzaman, K.J. Ansari, M. Mursaleen, On the parametric approximation results of Phillips operators involving the q-Appell polynomials. Iranian J. Sci. Technol. Trans.: Sci. 1–13 (2022)
27. Md. Nasiruzzaman, K.J. Ansari, M. Mursaleen, Weighted and voronovskaja type approximation by q-Szász-Kantorovich operators involving Appell polynomials. Filomat **37**, 67–84 (2023)
28. N. Rao, M. Ayman-Mursaleen, R. Aslan, A note on a general sequence of $\lambda$-Szász Kantorovich type operators. Comput. Appl. Math. **43**, 428 (2024)
29. N. Rao, M. Farid, M. Raiz, Symmetric properties of $\lambda$-Szász operators coupled with generalized beta functions and approximation theory. Symmetry **16** 1703 (2024)
30. Q.-B. Cai, R. Aslan, F. Özger, H.M. Srivastava, Approximation by a new Stancu variant of generalized $(\lambda, \mu)$-Bernstein operators. Alex. Eng. J. **107**, 205–214 (2024)
31. R.A. DeVore, G.G. Lorentz, *Constructive Approximation* (Springer, Berlin, Heidelberg, Germany, 1993)
32. F. Altomare, M. Campiti, Korovkin-type approximation theory and its applications. Walter de Gruyter, Berlin, Germany, vol. 17, (2011)
33. S.A. Mohiuddine, A. Alotaibi, B. Hazarika, Weighted $A$-statistical convergence for sequences of positive linear operators. Sci. World J. 437863 (2014)
34. S.A. Mohiuddine, Statistical weighted A-summability with application to Korovkin's type approximation theorem. J. Inequal. Appl. **2016**, 101 (2016)
35. U. Kadak, Statistical summability of double sequences by the weighted mean and associated approximation results, in *Applied Mathematical Analysis: Theory, Methods, and Applications* (Springer, Cham. 2020), pp. 61–85
36. R. Aslan, A. İzgi, Approximation by one and two variables of the Bernstein-Schurer-type operators and associated GBS operators on symmetrical mobile interval. J. Funct. Spaces *2021*, 1–12 (2021)
37. L.T. Su, R. Aslan, F.S. Zheng, M. Mursaleen, On the Durrmeyer variant of q-Bernstein operators based on the shape parameter $\lambda$. J. Inequal. Appl. **1**, 1–17 (2023)
38. H.M. Srivastava, K. Ansari, F. Özger, Z. Ödemis Özger, A link between approximation theory and summability methods via four-dimensional infinite matrices. Mathematics **2021**, *9* (1895). https://doi.org/10.3390/math9161895
39. F. Özger, K.J. Ansari, Statistical convergence of bivariate generalized Bernstein operators via four-dimensional infinite matrices. Filomat *36*, 507–525 (2022). https://doi.org/10.2298/FIL2202507O
40. Q.B. Cai, R. Aslan, On a new construction of generalized $q$-Bernstein polynomials based on shape parameter $\lambda$. Symmetry **2021**, 13 (1919)

41. R. Aslan, Approximation by Szasz-Mirakjan-Durrmeyer operators based on shape parameter $\lambda$. Commun. Fac. Sci. Univ. Ank. Ser. A1 Math. Stat. *71*, 407–421 (2022)
42. U. Kadak, F. Özger, A numerical comparative study of generalized Bernstein-Kantorovich operators. Math. Found. Comput. **4**, 311 (2021)
43. Q.B. Cai, K.J. Ansari, M. Temizer Ersoy, F. Özger, Statistical blending-type approximation by a class of operators that includes shape parameters $\lambda$ and $\alpha$. Mathematics *10*, 1149 (2022)
44. F. Özger, Applications of generalized weighted statistical convergence to approximation theorems for functions of one and two variables. Numer. Funct. Anal. Optim. **41**, 1990–2006 (2020). https://doi.org/10.1080/01630563.2020.1868503
45. K.J. Ansari, F. Özger, Z. Ödemis Özger, Numerical and theoretical approximation results for Schurer–Stancu operators with shape parameter $\lambda$ . Comput. Appl. Math. *41*, 1–18 (2022)
46. F. Özger, Weighted statistical approximation properties of univariate and bivariate $\lambda$-Kantorovich operators. Filomat **33**, 11 (2019)
47. F. Özger, On New Bezier Bases with Schurer Polnomials and corresponding results in approximation theory. Commun. Fac. Sci. Univ. Ank. Ser. A1 Math. Stat. **69**, 376–393 (2020)
48. F. Özger, E. Aljimi, M. Temizer Ersoy, Rate of weighted statistical convergence for generalized blending-type Bernstein-Kantorovich operators. Mathematics **2022**, 10 (2027)
49. U. Kadak, N.L. Braha, H.M. Srivastava, Statistical weighted B-summability and its applications to approximation theorems. Appl. Math. Comput. **302**, 80–96 (2017)

# Convergence Results for Nonlinear Generalized Bernstein Type Operators

**Sahil Berwal and Arun Kajla**

**Abstract** In this chapter we construct a new sequence of generalized nonlinear Bernstein-type operators. For these operators, we prove a Kelisky-Rivlin type theorem and determine their convergence rate. A fixed point type theorem and a generalized JS-contraction mapping are introduced.

**Keywords** JS-contraction · Nonlinear operators · Rate of convergence · Fixed point theorem · Contraction mapping

## 1 Introduction

Kelisky et al. [27] studied the convergence of the iterates of the Bernstein polynomial $B_k$ as $k \to \infty$, both when $k$ is a function of $n$ and when $k$ is independent of $n$, in 1967. Currently, the proof is recognized by the Kelisky-Rivlin Theorem. Two fundamental functional-type identities are established by German Badıa et al. [1] between the iterates of the Bleimann-Butzer-Hahn operator and the Bernstein operator. Certain Stancu operators have been identified to be weakly Picard operators by Rus [36], and [17] yielded similar results. Rus [37] provided a straightforward proof of the Kelisky-Rivlin Theorem using a contraction principle method. Jachymski provided a different proof in [19], which is predicated on a fixed point theorem for linear operators on a Banach space. A theorem for assessing whether infinite products of a linear operator would converge on an arbitrary Banach space was developed by Łukawska et al. [18] and derive a new version of the Kelisky-Rivlin theorem for Bernstein operator iterates in C[0, 1].

Mahmudov [29] developed a Korovkin type theorem for iterates of general positive linear operators and have shown that the iterates converge to a fixed point of

S. Berwal · A. Kajla (✉)
Department of Mathematics, Central University of Haryana, Mahendragarh, Haryana, India
e-mail: rachitkajla47@gmail.com

S. Berwal
e-mail: sahil211939@cuh.ac.in



S. A. Mohiuddine et al. (eds.), *Positive Operators and Fixed-Point Theorems with Applications*, Industrial and Applied Mathematics,
https://doi.org/10.1007/978-981-96-2476-8_2

the original operator under certain natural conditions. The results hold for real q-Bernstein-Durrmeyer operators, q-Stancu operators, q-MKZ operators, and Cesaro operators. Gavrea et al. [16] provide a new approach to studying the asymptotic behavior of its iterates of positive linear operators. The class of operators for which the limit of its iterates can be found is expanded by this method. Catinas [13] investigates the convergence of a few bivariate and trivariate Cheney-Sharma operators using the weakly Picard operators approach. He generalizes the process for the multivariate case as well. Wang [39] investigated the iterated q-Bernstein-Stancu operator properties.

Oruç and Tuncer demonstrated the convergence of the q-Bernstein operator iterates in [32]. According to Radu [34], the limit behavior of the iterates of the q- and (p,q)-Bernstein operators aims to prove convergence from some different viewpoints, such as using the contraction principle (Weakly Picard Operators Theory) in order to prove convergence. The theory of weakly Picard operators is widely applicable to those equations for which the successive approximation method is effective. Modified $(p, q)$-Bernstein-Schurer operators are introduced by Mursaleen et al. [30], who also examine their statistical approximation qualities using the Korovkins type approximation theorem. Through the application of the contraction principle, Rahman et al. [35] prove a Kelisky-Rivlin type theorem on the power of the q-Bernstein operators for two-dimensional cases, $(p, q)$-Bernstein operators, and bivariate $(p, q)$-Bernstein operators. Jebreen et al. [12] address the limit behavior of iterates of Lupaş $q$- and $(p, q)$-Bernstein operators and use the contraction principle to determine the convergence for Lupaş $q$- and $(p, q)$-Bernstein operators.

The definition of the Reich multivalued monotone contraction maps is discussed by Alfuraidan et al. [7]. In addition, they present a vector-valued Bernstein operator on the space C([0, 1], X), where X is a partial order Banach space. Next, they present a Kelisky-Rivlin theorem analog. A fixed point theorem was established by Altun et al. [6] for a novel class of mappings of the JS-contraction type. As applications, Kelisky-Rivlin type for both linear and nonlinear q-Bernstein-Stancu operators is developed. Some existence theorems for coupled fixed points of a Θ-type contractive operator in metric spaces equipped with a directed graph are established by Aslam et al. [10]. Pritha [33] presented a novel proving technique that allows the proof of the fixed point theorem for JSC-contraction in group theory. Mahloul et al. [28] prove a fixed point theorem that is novel and unique in the existing literature on fixed point theory, as well as develop a Kelisky-Rivlin type theorem for $\alpha$-Bernstein-type operators.

## 2 Preliminaries

Let $Y \neq \phi$ be a set and let $f : Y \to Y$ be a mapping. The Picard sequence is denoted by the sequence $\{f^m x_0\}_{m\in\mathbb{N}}$. This sequence is defined by the equation $x_m = f x_{m-1} = f^m x_0$ for all $m \in \mathbb{N}$, where $f_m$ is the result of repeatedly applying function $f$ for $m$ times. When $m = 0$, it acts as the identity function $I_y$. The recursive formula for $f^m$ is $f^m = f(f^{m-1})$, $m \in \mathbb{N}$.

In Banach's thesis [11], the fixed point theorem of an explicit version of the Banach contraction principle was utilized to demonstrate the existence of a solution to an integral equation. If $(Y, d)$ is a complete metric space and $L : Y \to Y$ is a contraction map (i.e., $d(Lx, Ly) \leq \lambda d(x, y)$ for all $x, y \in Y$, where $\lambda \in (0, 1)$ is a constant), then $L$ has a unique fixed point according to this principle. A novel fixed point theorem and a new class of contraction maps are proposed by Jleli and Samet [20], based on the setting of generalized metric spaces that we discuss in section three. In [6], a fixed point theorem for new classes of JS-contractions mappings was developed by Altun et al. generalizing the work of Jleli and Samet. As an application of Theorem 2.5, they also looked at the iterative property for linear and nonlinear, q-Bernstein-Stancu type operators.

Assuming $A$ is a group under addition, if we have a mapping $f$ on the set $A$, then the fixed points of function $f$ (Fix($f$)) are the elements that don't change when we apply the mapping $f$ to them. In other words, Fix(f) is the set of all elements $y \in A$ such that $f(y) = y$.

If $y \in B$, and $B \subseteq A$ then

$$y + B = y + x, \qquad x \in B.$$

The Bernstein basis polynomials of degree $m$ are defined as

$$B_m(\zeta) = \sum_{j=0}^{m} \binom{m}{j} y^j (1-y)^{m-j} \zeta\left(\frac{j}{m}\right), \qquad m \in \mathbb{N} \text{ and } \zeta \in C[0, 1]. \tag{2.1}$$

The sequence of iterates $(B_m^n)_{m \in \mathbb{N}}$ of $B_m$ is defined as

$$(B_m^n \zeta)(y) = [B_m(B_m^{n-1}\zeta)](y), \qquad \text{for} \quad m \in \mathbb{N}.$$

The following convergence result is proved by Kelisky-Rivlin [27].

**Theorem 2.1** ([27], Theorem 1)
*For any $\zeta \in C[0, 1]$, for any fixed $m \in \mathbb{N}$,*

$$\lim_{n\to\infty} (B_m^n \zeta)(y) = \zeta(0) + [\zeta(1) - \zeta(0)]y, \qquad y \in [0, 1].$$

*Later, Agratini and Rus [3] provided an extension of Theorem 1 in the following manner.*

**Theorem 2.2** ( [3], Theorem 4)
*Let $L_m : C[c, d] \to C[c, d]$ be defined as follows for each $m \in \mathbb{N}$, then*

$$(L_m f)(y) = \sum_{i=0}^{m} \psi_{m,i}(y) f(y_{m,i})$$

*where* $c = y_{m,0} < y_{m,1} < \cdots < y_{m,m} = d$ *and for every* $0 \leq i \leq m$*, the following conditions are true:*

$$\psi_{m,i} \in C[c,d], \quad \psi_{m,i} \geq 0 \quad \text{and} \quad \psi_{m,0}(c) = \psi_{m,m}(d) = 1.$$

*Let's use the following notation* $u_m := \min_{y \in [c,d]} (\psi_{m,0}(y) + \psi_{m,m}(y))$*. If the value of* $u_m > 0$*, then the sequence of iterates* $(L_m^n)_{n \geq 1}$ *converges uniformly on the interval* $[c, d]$ *and*

$$\lim_{n\to\infty} (L_m^n f)(x) = f(c) + \frac{f(d) - f(c)}{(d - c)}(x - c), \qquad f \in C[c,d].$$

*By Banach's contraction principle, Rus [11, 37] proposed the proof of Theorem 2.1.*

*A weakly Picard operator (WPO) on a Banach space* $Y$ *is mapping* $L : Y \to Y$*, such that the limit of* $L$ *is a fixed point and the iterative sequence* $(L^m(y))_{m \geq 1}$ *converges for all* $y \in Y$*.*

*Jachymski [19] extended Theorem 2.1 in the following way.*

**Theorem 2.3** ([19], Theorem 4.1) *Let* $L : Y \to Y$ *be a linear operator such that* $\|L|_{Y_0}\| \leq 1$*, where* $Y_0$ *which is closed subspace of a Banach space* $Y$*. If the range of the corresponding operator* $I_Y - L$ *is contained within* $Y_0$*, then* $L$ *is called Weakly Picard Operator (WPO).*

*Moreover,*

$$card Fix(L) = card\,\{y + Y_0, y \in Y\},$$

*and*

$$(y + Y_0) \cap Fix(L) = \left\{\lim_{n\to\infty} L^n y\right\}, \quad \text{for} \quad y \in Y.$$

*Sultana and Vetrivel [38] presented a generalization and enhancement of Theorem 2.3. Through their research, they proved that the linearity assumption of operator L is unnecessary. Consequently, for the nonlinear Bernstein operators, they developed a Kelisky-Rivlin type theorem and the operators are defined as follows:*

$$(\hat{B}_m \zeta)(y) = \sum_{k=0}^{m} \left|\zeta\left(\frac{k}{m}\right)\right| \binom{m}{k} y^k (1-y)^{m-k}, \text{ for } m \in \mathbb{N},\ y \in [0,1]. \tag{2.2}$$

*The conclusion in [38] addresses convergence of the sequence of* $\hat{B}_m$ *iterates as follows.*

**Theorem 2.4** ([38], Corollary 4) *Consider the set* $Y' = \{\xi \in C([0,1]); \xi(0) \geq 0$ *and* $\xi(1) \geq 0\}$*. For any continuous function* $g \in C[0,1]$*, the sequence* $(\hat{B}_m^j g)_{j \in \mathbb{N}}$ *converges to a fixed point of the operator* $\hat{B}_m$*.*

*Furthermore, for any $g \in Y'$, the following statement holds:*

$$\lim_{j\to\infty}\left(\hat{B}_m^j g\right)(y) = g(0) + [g(1) - g(0)]y.$$

*From this point forward, E refers to a group in relation to a "+" operation. Additionally, we indicate "$O_E$" as identity element of E.*

**Theorem 2.5** ([6], Corollary 2.2) *Consider a closed subset $Y_0 \subset Y$, where $Y_0$ represents a subgroup of E. Consider Y as a subset of E that possesses metric d and the metric space (Y, d) is complete. If $L : Y \to Y$ then the following statement is true:*

$$(y, z) \in Y \times Y, \qquad y - z \in Y_0, \qquad d(Ly, Lz) \leq kd(y, z),$$

*a constant k ∈(0,1) and the operation mapping $\pm : Y \times Y \to Y$ defined by*

$$\pm\,(z, y) = z \pm y, \qquad (z, y) \in (Y \times Y).$$

*Moreover, operation mapping is continuous in relation to d and*

$$y - Ly \in Y_0,\ y \in Y,$$

*then we have*

*(i) The operator L possesses a fixed point denoted as $\upsilon$, and for every $y_0 \in Y$, the Picard sequence $L^n y_0 \to \upsilon$.*

*(ii) For every $y_0 \in Y$, we get*

$$(y_0 + Y_0) \cap Fix(L) = \left\{\upsilon = \lim_{n\to\infty} L^n y_0\right\}.$$

This work is organized as follows: In the second section, we introduce a new family of operators known as generalized nonlinear Bernstein type operators. The third section is dedicated to studying significant results related to JS-contraction type mapping and associated definitions. Moving on to the fourth section, we present the proof of the Kelisky-Rivlin type theorem. Finally, we wrap up the chapter with a conclusion.

## 3 Construction of Nonlinear Operator

A generalization of the Bernstein operators with a shape parameter $\wp \in [0, 1]$ was presented by Chen et al. [14]. It is defined as

$$B_\rho^{(\wp)}(\zeta; y) = \sum_{i=0}^{\rho} p_{\rho,i}^{(\wp)}(y)\zeta\left(\frac{i}{\rho}\right),\ y \in \mathcal{J}, \zeta \in C(\mathcal{J}), \mathcal{J} = [0, 1], \qquad (3.1)$$

$$p_{\rho,i}^{(\wp)}(y)=\left[\binom{\rho-2}{i}(1-\wp)y+\binom{\rho-2}{i-2}(1-\wp)(1-y)+\binom{\rho}{i}\wp y(1-y)\right]y^{i-1}(1-y)^{\rho-i-1},$$

with $\rho\geq 2$, $p_{1,0}^{(\wp)}(y)=(1-y)$, $p_{1,1}^{(\wp)}(y)=y$.

It is important to note that these operators are only applicable to continuous functions. In these operators, $\wp$ is the shape parameter, which is used to adjust the shape of the approximation curve that is generated by the operator by changing the value of $\wp$ on the range [0, 1]. Khursheed et al. [9] introduced bivariate Bernstein-Kantorovich type operators based on multiple shape parameters. They also shared another set called Schurer-Stancu operators [8], featuring a shape parameter denoted by $\lambda$. For more information, the reader can read [4, 5, 21–26, 31, 40].

Acu and Kajla [2] presented the generalized Bernstein-type operators using parameters $r_1, r_2\in\mathbb{N}\cup\{0\}$, for every $\rho-r_1r_2\geq 1$, which is as follows:

$$\mathcal{B}_{\rho,r_1,r_2}^{(\wp)}(\zeta;y)=\sum_{\mu=0}^{\rho-r_1r_2}p_{\rho-r_1r_2,\mu}^{(\wp)}(y)\sum_{k=0}^{r_2}p_{r_2,k}^{(\wp)}(y)\zeta\left(\frac{\mu+kr_1}{\rho}\right), \tag{3.2}$$

where $p_{0,0}^{(\wp)}(y)=1$. If $r_1=r_2=0$, then the above operators reduce to (3.1).

We define a nonlinear generalized Bernstein-type operators as follows:

$$\hat{\mathcal{B}}_{\rho,r_1,r_2}^{(\wp)}(\zeta;y)=\sum_{\mu=0}^{\rho-r_1r_2}p_{\rho-r_1r_2,\mu}^{(\wp)}(y)\sum_{k=0}^{r_2}\left|\zeta\left(\frac{\mu+kr_1}{\rho}\right)\right|p_{r_2,k}^{(\wp)}(y). \tag{3.3}$$

For $r_1=r_2=0$, in (3.3), Eq. (3.3) reduces to the nonlinear $\wp$-Bernstein operators and also we take $\wp=1$ in Eq. (3.3), and we derive the Bernstein nonlinear operators (2.2).

## 4 Mapping JS-Contraction Types

To begin with, we will go over some well-established definitions and outcomes that will be necessary for the rest of this discussion. Jleli and Samet [20] introduced the idea of JS-contraction mappings, and their definition is presented as follows (also refer to Refs. [18, 19]).

**Definition 4.1** In the metric space $(Y,d)$, a mapping $L:Y\to Y$ is called a JS-contraction if there exist values $\theta$ belonging to some set $\Phi$, and a constant $0<i<1$, such that for all elements $y$ and $z$ in the space $Y$

$$d(Ly,Lz)>0\Rightarrow\theta(d(Ly,Lz))\leq[\theta(d(y,z))]^{i}.$$

In this context, the symbol $\Phi$ represents a set of functions denoted as $\theta:(0,\infty)\to(0,\infty)$. These functions must satisfy certain specified conditions:

$(\theta_1)$ $\theta$ is non-decreasing.
$(\theta_2)$ For every sequence $\{t_m\} \subset (0, \infty)$, the following is true:

$$\lim_{m \to \infty} \theta\,(t_m) = 1 \Longleftrightarrow \lim_{m \to \infty} t_m = 0^+.$$

$(\theta_3)$ There exist $r$ a real number between 0 and 1, and $l$ a positive real number such that

$$\lim_{t \to 0^+} \frac{\theta(t) - 1}{t^r} = l.$$

The research paper referenced as [20] presented the following outcome related to fixed points.

**Theorem 4.1** *Let $(Y, d)$ be the complete metric space, if $L : Y \to Y$ is a JS-contraction mapping, then L possesses one specific fixed point.*

*Remark 4.1* [11] If we consider the JS-contraction mapping $L : Y \to Y$ in the previous theorem and choose $\theta(t) = e^{\sqrt{t}}$, then the theorem transforms into a well-known result known as the Banach contraction mapping principle.

In a recent paper, Altun et al. [15] proved the next fixed point theorem by grouping together JS-contraction kind mappings.

**Theorem 4.2** ([15], Theorem 2.1)

*Consider a complete metric space $(Y, d)$, where a closed subset $Y_0$ of $Y$ forms a subgroup of a set E. Let Y be a subset of E equipped with a specific metric d. Suppose $L : Y \to Y$ is a mapping satisfying*

$$(y, z) \in Y \times Y, \quad y - z \in Y_0, \quad d(Ly, Lz) > 0 \Rightarrow \theta(d(Ly, Lz)) \leq [\theta(d(z, y))]^k. \tag{4.1}$$

*Given that $k \in (0, 1)$ represents a constant, and $\theta \in \Phi$, the operation mapping $\pm : Y \times Y \to Y$ is defined as follows:*

$$\pm\,(z, y) = z \pm y, \quad (y, z) \in Y \times Y.$$

*Moreover, operation mapping is continuous in relation to d and*

$$z - Lz \in Y_0, \quad z \in Y.$$

*Subsequently, we possess the following:*
*(i) L possesses a fixed point $\upsilon$, and for every value $z_0 \in Y$, the Picard sequence $\{L^m z_0\}$ approaches $\upsilon$ in its convergence.*
*(ii) For every $z_0 \in Y$, we have*

$$(z_0 + Y_0) \cap \text{Fix}(L) = \left\{\upsilon = \lim_{m \to \infty} L^m z_0\right\}.$$

*Let $\mathcal{S}$ denote the family of functions such that $\zeta : (0, \infty) \to (0, 1) : \lim_{t \to a^+}$ $(sup\zeta(t)) < 1, \quad \forall a > 0.$*

**Definition 4.2** Let $Y_0 \subset Y$ such that $Y_0$ is a subgroup of $E$ and $Y \subset E$ equipped with a specific metric d such that $(Y, d)$ is complete. Suppose that a generalized JS-contraction type function is $L : Y \to Y$, $\beta \in \mathcal{S}$ and a non-decreasing function $\theta : (0, \infty) \to (1, \infty)$ exist such that for any $(y, z) \in Y \times Y$ with $y - z \in Y_0$, then

$$d(Ly, Lz) > 0 \Rightarrow \theta(d(Ly, Lz)) \leq [\theta(d(z, y))]^{\beta(d(z,y))}. \tag{4.2}$$

*Remark 4.2* Due to the non-decreasing characteristic of the function $\theta$, the contraction condition holds true for a generalized JS-contraction type mapping $L$. Then

$$d(Ly, Lz) < d(y, z), \tag{4.3}$$

for all $(y, z) \in Y \times Y$ with $y - z \in Y_0$ and $Ly \neq Lz$.

Using the contraction inequality (6) and weaker assumptions, Mahloul et al. [28] improves and generalizes Theorem 4.2.

**Theorem 4.3** ([28]) *Consider a closed subset of $Y_0$ of $Y$ such that $Y_0$ forms a subgroup of $E$ in the metric space $(Y, d)$, where $Y \subset E$ is supplied with a certain distance metric making $(Y, d)$ complete. Let's have a generalized JS-contraction type mapping $L : Y \to Y$. If we use the following definition for the operation mapping $\pm : Y \times Y \to Y$*

$$\pm (y, z) = y \pm z, \quad (y, z) \in Y \times Y.$$

*With regard to the metric d, operation mapping is continuous:*

$$z - Lz \in Y_0, \quad z \in Y. \tag{4.4}$$

*Then we have*

*i. The mapping $L$ possesses a fixed point $\upsilon$, and for any given $z_0 \in Y$, the Picard sequence $\{L^n z_0\}$ converges toward $\upsilon$.*

*ii. For every $z_0 \in Y$,*

$$(z_0 + Y_0) \cap \mathrm{Fix}(L) = \left\{\upsilon = \lim_{n\to\infty} L^n z_0\right\}.$$

*Remark 4.3* When we set $\beta(t) = k$, where $k \in (0, 1)$, in Theorem 4.3, we recover Theorem 4.2. This means that Theorem 4.2 can be considered as a special case of Theorem 4.3 by choosing a specific value for the parameter $\beta(t)$.

*Remark 4.4* Theorem 4.3 provides a generalization and improvement over Theorem 4.2 in two aspects. Firstly, in Theorem 4.2, the constant $k \in (0, 1)$ has been replaced with a function in Theorem 4.3, making the result more flexible and applicable in a broader range of situations. Secondly, Theorem 4.3 omits both conditions $(\Phi_2)$ and $(\Phi_3)$ that were required in Theorem 4.2, which means the assumptions for Theorem 4.3 are weaker, leading to more relaxed conditions for the theorem to hold.

Mahloul et al. [28] obtain the following Corollary taking $\theta(t) = e^t$ in Theorem 4.3.

**Corollary 4.1** *Consider $Y_0 \subset Y$ as a closed subset of $Y$, satisfying the condition that $Y_0$ is a subgroup of $E$, and let $Y \subset E$ equipped with a specific metric $d$ such that $(Y, d)$ is complete. Let $T$ be a self-mapping on $Y$ such that*

$$\begin{aligned}&(y, z) \in Y \times Y, \quad y - z \in Y_0,\\ &d(Ly, Lz) > 0 \Rightarrow d(Ly, Lz) \leq \beta(d(y, z))d(y, x),\end{aligned}$$

*where $\beta \in \mathcal{S}$. Suppose the $\pm : Y \times Y \to Y$ defined by*

$$\pm\,(y, z) = y \pm z, \quad (y, z) \in Y \times Y.$$

*Using the metric $d$, operation mapping is continuous:*

$$y - Ly \in Y_0, \quad y \in Y.$$

*Then we have*
*(i) The mapping $L$ has a fixed point $\upsilon$, and for any given $y_0 \in Y$, the Picard sequence $\{L^n y_0\}$ converges to $\upsilon$.*
*(ii) For every $y_0 \in Y$,*

$$(y_0 + Y_0) \cap \mathrm{Fix}(L) = \left\{\upsilon = \lim_{n\to\infty} L^n y_0\right\}.$$

*Remark 4.5* Using $k \in (0, 1)$ and putting $\beta(t) = k$ into Corollary 4.1, we obtain Theorem 2.5.

**Corollary 4.2** *In the context of a norm space $E$ and a Banach space $Y$ such that $Y_0$ is a closed subspace of $Y$, consider a mapping $L : Y \to Y$. If there exists a function $\beta \in \mathcal{S}$ such that for all $(y, z) \in Y \times Y$, the condition $|Ly - Lx| \leq \beta(|y - z|)|y - z|$ holds whenever $y - z \in Y_0$ and $Ly \neq Lz$, and if $(I_Y - L)(Y) \subseteq Y_0$, then the following statements are true:*
*(i) The mapping $L$ possesses a fixed point $\upsilon$, and for each $z_0 \in Y$, the Picard sequence $\{L^m z_0\}$ converges to $\upsilon$.*
*(ii) For every $z_0 \in Y$,*

$$(z_0 + Y_0) \cap \mathrm{Fix}(L) = \left\{\upsilon = \lim_{m\to\infty} L^m z_0\right\}.$$

***Proof*** The result can be derived by considering the distance function $d(y, z) = |y - z|$ in Corollary 4.1.

*Remark 4.6* Using set $E = Y$ in Corollary 4.2, we get the result presented in ([38], Theorem 6).

## 5 Convergence of Operators

We provide a proof for the convergence of the iterative sequence of nonlinear generalized Bernstein-type operators in this section, which is defined in Eq. (3.3). The key to establishing this convergence lies in utilizing the fixed point result found in the preceding section, specifically, Corollary 4.1 in Sect. 2.

To prove our convergence theorem, we begin by recalling some fundamental properties of $\mathcal{B}^{\wp}_{\rho,r_1,r_2}(\zeta; y)$ from [2]. These properties will be crucial in our subsequent analysis.

For $\zeta \in C(\jmath)$, we present the generalized Bernstein-type operators using parameters $r_1, r_2 \in \mathbb{N} \cup 0$, for every $\rho - r_1 r_2 \geq 1$ and for fixed real number as follows:

$$\mathcal{B}^{\wp}_{\rho,r_1,r_2}(\zeta; y) = \sum_{\mu=0}^{\rho-r_1r_2} p^{\wp}_{\rho-r_1r_2,\mu}(y) \sum_{k=0}^{r_2} p^{\wp}_{r_2,k}(y)\zeta\left(\frac{\mu + kr_1}{\rho}\right), \tag{5.1}$$

where

$$p^{\wp}_{\rho,k}(y) = \left[\binom{\rho-2}{k}(1-\wp)y + \binom{\rho-2}{k-2}(1-y)(1-\wp) + \binom{\rho}{k}\wp y(1-y)\right](1-y)^{\rho-k-1}y^{k-1}, \tag{5.2}$$

$$\sum_{k=0}^{\rho} p^{\wp}_{\rho,k}(y) = 1, \tag{5.3}$$

(a) $p^{\wp}_{1,0}(y) = (1-y)$.
(b) $p^{\wp}_{0,0}(y) = 1$.
(c) $p^{\wp}_{1,1}(y) = y$.

Nonlinear generalized Bernstein-type operators using parameters $r_1, r_2 \in \mathbb{N} \cup \{0\}$,

$$\hat{\mathcal{B}}^{\wp}_{\rho,r_1,r_2} = \sum_{\mu=0}^{\rho-r_1r_2} p^{\wp}_{\rho-r_1r_2,\mu}(y) \sum_{k=0}^{r_2} p^{\wp}_{r_2,k}(y)\left|\zeta\left(\frac{\mu + kr_1}{\rho}\right)\right|. \tag{5.4}$$

**Lemma 5.1** ([2]) *Let* $e_i(y) = y^p$, $p = \overline{0, 1}$. *We have*
*(i)* $\mathcal{B}^{(\wp)}_{\rho,r_1,r_2}(e_0; y) = 1$.
*(ii)*$\mathcal{B}^{\wp}_{\rho,r_1,r_2}(y; y) = y$.

**Lemma 5.2** $\mathcal{B}^{(\wp)}_{\rho,r_1,r_2}$ *operators posses end point interpolation property, where*
*(i)* $\mathcal{B}^{\wp}_{\rho,r_1,r_2}(\zeta; 0) = \zeta(0)$.
*(ii)* $\mathcal{B}^{\wp}_{\rho,r_1,r_2}(\zeta; 1) = \zeta(1)$.

***Proof*** (i) We have

$$\mathcal{B}^{\wp}_{\rho,r_1,r_2}(\zeta;0)=\sum_{\mu=0}^{\rho-r_1r_2}p^{\wp}_{\rho-r_1r_2,\mu}(0)\sum_{k=0}^{r_2}p^{\wp}_{r_2,k}(0)\zeta\left(\frac{\mu+kr_1}{\rho}\right),$$

$$\mathcal{B}^{\wp}_{\rho,r_1,r_2}(\zeta;0)=\sum_{\mu=0}^{\rho-r_1r_2}p^{\wp}_{\rho-r_1r_2,\mu}(0)\left[p^{\wp}_{r_2,0}(0)\zeta\left(\frac{\mu}{\rho}\right)+p^{\wp}_{r_2,1}(0)\zeta\left(\frac{\mu+r_1}{\rho}\right)+\cdots+p^{\wp}_{r_2,r_2}(0)\zeta\left(\frac{\mu+r_1r_2}{\rho}\right)\right].$$

Using equation (11) find the value of $p^{\wp}_{r_2,0}(0)=1$, $p^{\wp}_{r_2,1}(0)=0$, ..., $p^{\wp}_{r_2,r_2}(0)=0$; we get

$$\mathcal{B}^{\wp}_{\rho,r_1,r_2}(\zeta;0)=\sum_{\mu=0}^{\rho-r_1r_2}p^{\wp}_{\rho-r_1r_2,\mu}(0)\zeta\left(\frac{\mu}{\rho}\right).$$

Now,

$$\mathcal{B}^{\wp}_{\rho,r_1,r_2}(\zeta;0)=\zeta(0)\cdot p^{\wp}_{\rho-r_1r_2,0}(0)+\zeta\left(\frac{1}{\rho}\right)p^{\wp}_{\rho-r_1r_2,1}(0)+\cdots+\zeta\left(\frac{\rho-r_1r_2}{\rho}\right)p^{\wp}_{\rho-r_1r_2,\rho-r_1r_2}(0).$$

Using Eq. (5.2) we obtain value of

$$p^{\wp}_{\rho-r_1r_2,0}(0)=1,\ p^{\wp}_{\rho-r_1r_2,1}(0)=p^{\wp}_{\rho-r_1r_2,2}(0)=\cdots=p^{\wp}_{\rho-r_1r_2,\rho-r_1,r_2}(0)=0.$$

Now we get

$$\mathcal{B}^{\wp}_{\rho,r_1,r_2}(\zeta;0)=\zeta(0). \tag{5.5}$$

***Proof*** (ii) We have

$$\mathcal{B}^{\wp}_{\rho,r_1,r_2}(\zeta;1)=\sum_{\mu=0}^{\rho-r_1r_2}p^{\wp}_{\rho-r_1r_2,\mu}(1)\sum_{k=0}^{r_2}p^{\wp}_{r_2,k}(1)\zeta\left(\frac{\mu+kr_1}{\rho}\right),$$

$$\mathcal{B}^{\wp}_{\rho,r_1,r_2}(\zeta;1)=\sum_{\mu=0}^{\rho-r_1r_2}p^{\wp}_{\rho-r_1r_2,\mu}(1)\left[p^{\wp}_{r_2,0}(1)\zeta\left(\frac{\mu}{\rho}\right)+p^{\wp}_{r_2,1}(1)\zeta\left(\frac{\mu+r_1}{\rho}\right)+\cdots+p^{\wp}_{r_2,r_2}(1)\zeta\left(\frac{\mu+r_1r_2}{\rho}\right)\right].$$

Calculate the value of $p^{\wp}_{r_2,0}(1)$, $p^{\wp}_{r_2,1}(1)$, ..., $p^{\wp}_{r_2,r_2}(1)$; we get

$$p^{\wp}_{r_2,0}(1)=0,\ p^{\wp}_{r_2,1}(1)=0,\ ...,\ p^{\wp}_{r_2,r_2}(1)=1.$$

$$\mathcal{B}^{\wp}_{\rho,r_1,r_2}(\zeta;1)=\sum_{\mu=0}^{\rho-r_1r_2}p^{\wp}_{\rho-r_1r_2,\mu}(1)\zeta\left(\frac{\mu+r_1r_2}{\rho}\right),$$

$$\mathcal{B}^{\wp}_{\rho,r_1,r_2}(\zeta;1) = p^{\wp}_{\rho-r_1r_2,0}(1)\zeta(\frac{r_1r_2}{\rho}) + p^{\wp}_{\rho-r_1r_2,1}(1)\zeta\left(\frac{1+r_1r_2}{\rho}\right)$$
$$+\cdots+ p^{\wp}_{\rho-r_1r_2,\rho-r_1r_2}(1)\zeta\left(\frac{\rho+r_1r_2-r_1r_2}{\rho}\right).$$

Now using Eq. (5.2) find the value of

$$p^{\wp}_{\rho-r_1r_2,0}(1) = p^{\wp}_{\rho-r_1r_2,1}(1) = p^{\wp}_{\rho-r_1r_2,2}(1) = \cdots = p^{\wp}_{\rho-r_1r_2,\rho-r_1,r_2-1}(1) = 0,$$
$$p^{\wp}_{\rho-r_1r_2,\rho-r_1,r_2}(1) = 1.$$

Using the above relation, we get

$$\mathcal{B}^{\wp}_{\rho,r_1,r_2}(\zeta;1) = \zeta(1). \tag{5.6}$$

**Lemma 5.3** *Let* $\wp \in[0,1]$. *For all* $y\in[0,1]$, *we have*

$$p^{\wp}_{\rho-r_1r_2,0}(y)p^{\wp}_{r_2,0}(y) + p^{\wp}_{\rho-r_1r_2,\rho-r_1r_2}(y)p^{\wp}_{r_2,r_2}(y) \geq \frac{1}{2^{\rho-r_1r_2-r_2}}. \tag{5.7}$$

***Proof*** From Eq. (5.2), we have

$$\begin{aligned}p^{\wp}_{\rho-r_1r_2,0}(y)p^{\wp}_{r_2,0}(y) &= [(1-\wp)y+\wp y(1-y)]y^{-1}(1-y)^{\rho-r_1r_2-1}\\ &\quad [(1-\wp)y+\wp y(1-y)]y^{-1}(1-y)\\ &= (1-\wp)^2(1-y)^{\rho-r_1r_2+r_2-2} + 2\wp(1-\wp)(1-y)^{\rho-r_1r_2+r_2-1}\\ &\quad +\wp^2(1-y)^{\rho-r_1r_2+r_2}.\end{aligned}$$

$$\begin{aligned}p^{\wp}_{\rho-r_1r_2,\rho-r_1r_2}(y)p^{\wp}_{r_2,r_2}(y) &= [(1-y)(1-\wp)+(1-y)\wp y]\\ y^{\rho-r_1r_2-1}(1-y)^{-1}&[(1-\wp)(1-y)+\wp y(1-y)]y^{r_2-1}(1-y)^{-1}\\ &= (1-\wp)^2y^{\rho-r_1r_2+r_2-2} + 2\wp(1-\wp)y^{\rho-r_1r_2+r_2-1} + \wp^2y^{\rho-r_1r_2+r_2}.\end{aligned}$$

Using the above equation

$$\begin{aligned}&p^{\wp}_{\rho-r_1r_2,0}(y)p^{\wp}_{r_2,0}(y) + p^{\wp}_{\rho-r_1r_2,\rho-r_1r_2}(y)p^{\wp}_{r_2,r_2}(y)\\ &= (1-\wp)^2[(1-y)^{\rho-r_1r_2+r_2-2} + y^{\rho-r_1r_2+r_2-2}]\\ &\quad +2\wp(1-\wp)[(1-y)^{\rho-r_1r_2+r_2-1} + y^{\rho-r_1r_2+r_2-1}]\\ &\quad +\wp^2[(1-y)^{\rho-r_1r_2+r_2} + y^{\rho-r_1r_2+r_2}].\end{aligned}$$

From the following inequality:

$$(a^n+b^n) \geq \frac{(b+a)^n}{2^{n-1}} \qquad b,a\geq 0,$$

and the fact that $\wp \in[0,1]$, it follows that

$$\begin{aligned} & p^{\wp}_{\rho-r_1r_2,0}(y)p^{\wp}_{r_2,0}(y)+p^{\wp}_{\rho-r_1r_2,\rho-r_1r_2}(y)p^{\wp}_{r_2,r_2}(y) \\ \geq & \frac{(1-\wp)^2}{2^{\rho-r_1r_2+r_2-3}}+\frac{2\wp(1-\wp)}{2^{\rho-r_1r_2+r_2-2}}+\frac{\wp^2}{2^{\rho-r_1r_2+r_2-1}} \\ \geq & \frac{4(1-\wp)^2+4\wp(1-\wp)+\wp^2}{2^{\rho-r_1r_2+r_2-1}} \\ \geq & \frac{1}{2^{\rho-r_1r_2+r_2-1}}. \end{aligned}$$

For the nonlinear generalized Bernstein-type operators, we present a Kelisky-Rivlin type theorem using parameters $r_1$ and $r_2$. The theorem establishes conditions under which these operators achieve a certain degree of approximation or convergence to a function, given that $\rho - r_1r_2 \geq 1$.

**Theorem 5.1** *For every* $r_1, r_2 \in \mathbb{N} \cup \{0\}$ *and* $\wp \in$ *[0,1]. Then, for every* $\zeta \in (C[0, 1])$ *with* $\zeta(0)\geq 0$ *and* $\zeta(1) \geq 1$, *we have*

$$\lim_{m\to\infty} \hat{\mathcal{B}}^{m,\wp}_{\rho,r_1,r_2}(\zeta; y) = \zeta(0) + [\zeta(1) - \zeta(0)]y, \qquad y \in [0, 1].$$

***Proof*** Let $E = C[0, 1]$, and consider $Y \subset E$ defined by

$$Y = \{\zeta \in E : \zeta(1) \geq 0, \zeta(0) \geq (0)\}.$$

Now, the metric $d : Y \times Y \to [0, \infty)$, which is defined through

$$d(\zeta, \phi) = \sup|\zeta(y) - \phi(y)| \text{ for all } \zeta, \phi \in Y \text{ and } y \in [0, 1].$$

Clearly, $(Y, d)$ forms a complete metric space. Let $Y_0$ be a subset of $Y$ defined as

$$Y_0 = \zeta \in E : \zeta(0) = \zeta(1) = 0.$$

It is clear that $Y_0$ is a subgroup of $E$ and a closed subset of $Y$. The proof will be divided into several cases.

**Case 1**:

If $\rho - r_1r_2 = 1; \qquad r_1 = r_2 = 0,$

$$\begin{aligned} \hat{\mathcal{B}}^{\wp}_{1,0,0}(\zeta; y) &= \sum_{\mu=0}^{1} p^{\wp}_{1,\mu}(y) \sum_{k=0}^{0} p^{\wp}_{0,k}(y) \left|\zeta\left(\frac{\mu+k\times 0}{\rho}\right)\right| \\ &= \sum_{\mu=0}^{1} p^{\wp}_{1,\mu}(y)p^{\wp}_{0,0}(y) \left|\zeta\left(\frac{\mu}{\rho}\right)\right| \\ &= p^{\wp}_{1,0}(y)\,|\zeta\,(0)| + p^{\wp}_{1,1}(y)\,|\zeta\,(1)| \\ &= (1-y)\,|\zeta(0)| + y\,|\zeta(1)|. \end{aligned}$$

From the above equation, it follows that

$$\begin{aligned}\hat{\mathcal{B}}^{2,\wp}_{1,0,0}(\zeta; y) &= \left[\hat{\mathcal{B}}^{\wp}_{1,0,0}\left(\hat{\mathcal{B}}^{\wp}_{1,0,0}(\zeta; y); y\right)\right] \\ &= \left[\hat{\mathcal{B}}^{\wp}_{1,0,0}\left((1-y)\left|\zeta(0)\right| + y\left|\zeta(1)\right|; y\right)\right] \\ &= (1-y)\left|\zeta(0)\right| + y\left|\zeta(1)\right|.\end{aligned}$$

Inductively, we get

$$\lim_{m\to\infty} \hat{\mathcal{B}}^{m,\wp}_{1,0,0}(\zeta; y) = (1-y)\left|\zeta(0)\right| + y\left|\zeta(1)\right|, \qquad \forall m \in \mathbb{N}, \quad y \in [0,1].$$

**Case 2**: If $\rho - r_1 r_2 \geq 1$ and $r_1, r_2 \in \mathbb{N} \cup \{0\}$. Let $(\zeta, \phi) \in (\mathrm{Y} \times \mathrm{Y})$ be arbitrary such that $\hat{\mathcal{B}}^{\wp}_{\rho,r_1,r_2}(\zeta; y) \neq \hat{\mathcal{B}}^{\wp}_{\rho,r_1,r_2}(\phi; y)$ and $\zeta - \phi \in Y_0$. Note that

$$\zeta - \phi \in Y_0 \iff \zeta(0) = \phi(0), \zeta(1) = \phi(1).$$

Let $y \in [0,1]$ and $\wp \in [0,1]$. Then

$$\begin{aligned}&\left|\hat{\mathcal{B}}^{\wp}_{\rho,r_1,r_2}(\zeta; y) - \hat{\mathcal{B}}^{\wp}_{\rho,r_1,r_2}(\phi; y)\right| \\ &= \left|\sum_{\mu=0}^{\rho-r_1r_2} p^{\wp}_{\rho-r_1r_2,\mu}(y) \sum_{k=0}^{r_2} p^{\wp}_{r_2,k}(y)\left(\left|\zeta\left(\frac{\mu+kr_1}{\rho}\right)\right| - \left|\phi\left(\frac{\mu+kr_1}{\rho}\right)\right|\right)\right| \\ &= \left|p^{\wp}_{\rho-r_1r_2,0}(y)\sum_{k=1}^{r_2} p^{\wp}_{r_2,k}(y)\left|\zeta\left(\frac{kr_1}{\rho}\right) - \phi\left(\frac{kr_1}{\rho}\right)\right|\right. \\ &\quad + \sum_{\mu=1}^{\rho-r_1r_2-1} p^{\wp}_{\rho-r_1r_2-1,\mu}(y) \sum_{k=0}^{r_2} p^{\wp}_{r_2,k}(y)\left|\zeta\left(\frac{\mu+kr_1}{\rho}\right) - \phi\left(\frac{\mu+kr_1}{\rho}\right)\right| \\ &\quad \left.+ p^{\wp}_{\rho-r_1r_2,\rho-r_1r_2}(y)\sum_{k=0}^{r_2-1} p^{\wp}_{r_2,k}(y)\left|\zeta\left(\frac{\rho-r_1r_2+kr_1}{\rho}\right) - \phi\left(\frac{\rho-r_1r_2+kr_1}{\rho}\right)\right|\right| \\ &= \left|p^{\wp}_{\rho-r_1r_2,0}(y)\sum_{k=1}^{r_2} p^{\wp}_{r_2,k}(y)d(\zeta,\phi) + \sum_{\mu=1}^{\rho-r_1r_2-1} p^{\wp}_{\rho-r_1r_2-1,\mu}(y)\sum_{k=0}^{r_2} p^{\wp}_{r_2,k}(y)d(\zeta,\phi)\right. \\ &\quad \left.+ p^{\wp}_{\rho-r_1r_2,\rho-r_1r_2}(y)\sum_{k=0}^{r_2-1} p^{\wp}_{r_2,k}(y)d(\zeta,\phi)\right|\end{aligned}$$

$$
\begin{aligned}
&\leq d(\zeta,\phi)\left| p^{\wp}_{\rho-r_1r_2,0}(y)\sum_{k=1}^{r_2} p^{\wp}_{r_2,k}(y) + \sum_{\mu=1}^{\rho-r_1r_2-1} p^{\wp}_{\rho-r_1r_2-1,\mu}(y)\sum_{k=0}^{r_2} p^{\wp}_{r_2,k}(y) \right. \\
&\left. + p^{\wp}_{\rho-r_1r_2,\rho-r_1r_2}(y)\sum_{k=0}^{r_2-1} p^{\wp}_{r_2,k}(y) \right| \\
&\leq d(\zeta,\phi)\left(1 - p^{\wp}_{\rho-r_1r_2,0}(y)p^{\wp}_{r_2,0}(y) - p^{\wp}_{\rho-r_1r_2,\rho-r_1r_2}(y)p^{\wp}_{r_2,r_2}(y)\right).
\end{aligned}
$$

Using Eq. (5.7) we get

$$
\left|\hat{\mathcal{B}}^{\wp}_{\rho,r_1,r_2}(\zeta;y) - \hat{\mathcal{B}}^{\wp}_{\rho,r_1,r_2}(\phi;y)\right| \leq d(\zeta,\phi)\left(1 - \frac{1}{2^{\rho-r_1r_2+r_2-1}}\right).
$$

This implies that

$$
\left|\hat{\mathcal{B}}^{\wp}_{\rho,r_1,r_2}(\zeta;y) - \hat{\mathcal{B}}^{\wp}_{\rho,r_1,r_2}(\phi;y)\right| \leq \beta(d(\zeta,\phi))d(\zeta,\phi),
$$

where $\beta(t) = 1 - \dfrac{1}{2^{\rho-r_1r_2+r_2-1}}, \quad \forall\, t > 0.$

Considering the supremum in the inequality above for $y \in [0,1]$, we obtain

$$
d\left(\hat{\mathcal{B}}^{\wp}_{\rho,r_1,r_2}(\zeta), \hat{\mathcal{B}}^{\wp}_{\rho,r_1,r_2}(\phi)\right) \leq \beta(d(\zeta\phi))d(\zeta,\phi).
$$

Now, with each $\zeta \in X$, we put

$$
\chi(x) = \zeta(x) - \left(\hat{\mathcal{B}}^{\wp}_{\rho,r_1,r_2}(\zeta;x)\right).
$$

Using Lemma 5.2, we get

$$
\begin{aligned}
\chi(0) &= \zeta(0) - |\zeta(0)| \\
&= \zeta(0) - \zeta(0) = 0.
\end{aligned}
$$

$$
\begin{aligned}
\chi(1) &= \zeta(1) - |\zeta(1)| \\
&= \zeta(1) - \zeta(1) = 0.
\end{aligned}
$$

This implies

$$
\left(I_X - \hat{\mathcal{B}}^{\wp}_{\rho,r_1,r_2}(\zeta)\right)(Y) \subseteq Y_0.
$$

Then by Corollary 4.1, we have

$$
(\zeta + Y_0) \cap Fix\left(\hat{\mathcal{B}}^{\wp}_{\rho,r_1,r_2}\right) = \left\{\lim_{m\to\infty} \hat{\mathcal{B}}^{\wp^m}_{\rho,r_1,r_2}\zeta\right\},\ \zeta \in Y.
$$

Lemma 5.1 leads us to conclude that $e_i \in \text{Fix}\left(\hat{\mathcal{B}}^{\wp}_{\rho,r_1,r_2}\right)$, where $e_i = y^i$ for $i = 0, 1$, and $y \in [0, 1]$.

Since $\text{Fix}\left(\hat{\mathcal{B}}^{\wp}_{\rho,r_1,r_2}\right)$ is a subspace of $Y$, we can assume that $\lambda : [0, 1] \to \mathbb{R}$ is expressed as follows:

$$\lambda(y) = \zeta(0) + [\zeta(1) - \zeta(0)]\, y = \zeta(0)e_0 + [\zeta(1) - \zeta(0)]\, e_1,$$

which belongs to $\text{Fix}\left(\hat{\mathcal{B}}^{\wp}_{\rho,r_1,r_2}\right)$.

Also, the function $\xi(y) : [0,1] \to \mathbb{R}$ described by

$$\xi(y) = \lambda(y) - \zeta(y) = \zeta(0) + [\zeta(1) - \zeta(0)]\, y - \zeta(y)$$

satisfies $\xi(0) = 0, \xi(1) = 0$, Then $\xi \in Y_0$, therefore $\lambda \in \zeta + Y_0$.

As a consequence, we can conclude that

$$\lambda \in (\zeta + Y_0) \cap Fix\left(\hat{\mathcal{B}}^{\wp}_{\rho,r_1,r_2}\right) = \left\{\lim_{m\to\infty} \hat{\mathcal{B}}^{m,\wp}_{\rho,r_1,r_2}\zeta\right\}.$$

There for every $\zeta \in Y$, we get

$$\lim_{m\to\infty} \left\{\hat{\mathcal{B}}^{m,\wp}_{\rho,r_1,r_2}(\zeta; y)\right\} = \zeta(0) + [\zeta(1) - \zeta(0)]\, y,$$

which completes the proof of the theorem.

## 6 Conclusion

In this chapter, we introduced a new sequence of generalized nonlinear Bernstein-type operators. We established a Kelisky-Rivlin type theorem for these operators and analyzed their convergence rate. Additionally, we presented a fixed point theorem and a generalized JS-contraction mapping, further expanding the theoretical framework in this area. These results contribute to the study of nonlinear approximation and fixed point theory, providing potential directions for future research.

## References

1. J.A. Adell, F.G. Badıa, J. de la Cal, On the iterates of some Bernstein-type operators. J. Math. Anal. Appl. **209**(2), 529–541 (1997)
2. A.M. Acu, A. Kajla, Blending type approximation by modified Bernstein operators. Adv. Oper. Theory **7**, 9 (2022)
3. O. Agratini, I.A. Rus, Iterates of a class of discrete linear operators via contraction principle. Comment. Math. Univ. Carolin. **44**, 555–563 (2003)

4. P.N. Agrawal, A. Kajla, S. Singh, Modified $\alpha$-Bernstein-Durrmeyer-Type Operators. Iran. J. Sci. Technol. Trans A. Sci., 45(6), 2049-2061 (2021)
5. P.N. Agrawal, M. Goyal, A. Kajla, q-Bernstein-Schurer-Kantorovich type operators. Bolletino dell Unione Mat. Ital. **8**, 169–180 (2015)
6. I. Altun, A.N. Al, M. Jleli, A. Lashine, B. Samet, A fixed point theorem for JS-contraction type mappings with applications to polynomial approximations. Filomat **31**, 4969–4978 (2017)
7. M.R. Alfuraidan, M. Bachar, M.A. Khamsi, A graphical version of Reich's fixed point theorem. J. Nonlinear Sci. Appl. **9**, 3931–3938 (2016)
8. K.J. Ansari, F. Özger, Z. Ödemiş Özger, Numerical and theoretical approximation results for Schurer-Stancu operators with shape parameter $\lambda$. Comput. Appl. Math. **41**(4), 181 (2022)
9. K.J. Ansari, S. Karakılıç, F. Özger, Bivariate Bernstein-Kantorovich operators with a summability method and related GBS operators. Filomat **36**(19), 6751–6765 (2022)
10. Z. Aslam, J. Ahmad, N. Sultana, Computational coupled fixed points for $\theta-$contractive mappings in metric spaces endowed with a graph. Bull. Math. Anal. Appl. **10**(1), 89–99 (2018)
11. S. Banach, Sur les opérations dans les ensembles abstraits et leur application aux équations intégrales. [On the operations in abstract sets and their application to integral equations] Fund. Math. **3**, 133–181 (1922)
12. H. Bin Jebreen, M. Mursaleen, M. Ahasan, On the convergence of Lupaş (p, q)-Bernstein operators via contraction principle. J. Inequal. Appl. **2019**(1), 34 (2019)
13. T. Catinas, D. Otrocol, Iterates of multivariate Cheney-Sharma operators. J. Comput. Anal. Appl. **15**(7), 1240–1246 (2013)
14. X. Chen, J. Tan, Z. Liu, J. Xie, Approximation of functions by a new family of generalized Bernstein operators. J. Math. Anal. Appl. **450**, 244–261 (2017)
15. N. Çetin, V.A. Radu, Approximation by generalized Bernstein-Stancu operators. Turkish J. Math. **43**(4), 2032–2048 (2019)
16. I. Gavrea, M. Ivan, On the iterates of positive linear operators. J. Approx. Theory **163**(9), 1076–1079 (2011)
17. H. Gonska, P. Piţul, I. Raşa, Over-iterates of Bernstein-Stancu operators. Calcolo **44**, 117–125 (2007)
18. G. Gwóźdź-Łukawska, J. Jachymski, IFS on a metric space with a graph structure and extensions of the Kelisky-Rivlin theorem. J. Math. Anal. Appl. **356**(2), 453–463 (2009)
19. J. Jachymski, The contraction principle for mappings on a metric space with a graph. Proc. Am. Math. Soc. **136**, 1359–1373 (2008)
20. M. Jleli, B. Samet, A new generalization of the Banach contraction principle. J. Inequal. Appl. **2014**, 38 (2014)
21. U. Kadak, F. Özger, A numerical comparative study of generalized Bernstein-Kantorovich operators. Math. Found. Comput., 4(4), 311–332 (2021)
22. A. Kajla, A Kantorovich variant of a generalized Bernstein operators. J. Math. Comput. Sci. **19**, 86–96 (2019)
23. A. Kajla, T. Acar, Modified $\alpha$-Bernstein operators with better approximation properties. Ann. Funct. Anal. **10**(4), 570–582 (2019)
24. A. Kajla, Sahil, P. Sehrawat, Bézier-Bernstein-Schurer type operators. Gen. Math. **30**(2), 51–61 (2022)
25. A. Kajla, S. Berwal, P. Sehrawat, Approximation by Szász-integral type operators, Filomat 38(4), 1317–1327 (2024)
26. A. Kajla, F. Özger, J. Yadav, Bézier-Baskakov-Beta type operators. Filomat **36**(19), 6735–6750 (2022)
27. R. Kelisky, T. Rivlin, Iterates of Bernstein polynomials. Pacific J. Math. **21**(1967), 511–520 (1967)
28. N. Mahloul, H. Ramoul, M. Abbas, Convergence of iterates of $\alpha$-Bernstein type operators via fixed point of generalized JS-contraction type mappings. Numer. Funct. Anal. Optim. **43**, 580–598 (2022)
29. N.I. Mahmudov, Korovkin type theorem and iterates of certain positive linear operators (2010). arXiv:1012.1187

30. M. Mursaleen, A. Al-Abied, M. Nasiruzzaman, Modified (p, q)-Bernstein-Schurer operators and their approximation properties. Cogent Math. **3**(1), 1236534 (2016)
31. F. Özger, K.J. Ansari, Statistical convergence of bivariate generalized Bernstein operators via four-dimensional infinite matrices. Filomat **36**(2), 507–525 (2022)
32. H. Oruç, N. Tuncer, On the convergence and iterates of q-Bernstein polynomials. J. Approx. Theory **117**(2), 301–313 (2002)
33. A.J.S Pritha, A.M.P. Dharsini, A.L.M. Prakasam, A fixed point theorem for JSC-contraction type mappings on group theory. Eur. J. Mol. Clin. Med., 7(11), (2021)
34. V.A. Radu, Note on the iterates of q-and (p, q)-Bernstein operators. Sci. Stud. Res. Ser. Math. Inform **26**(2), 83–94 (2016)
35. S. Rahman, M. Mursaleen, A.H. Alkhaldi, Convergence of iterates of q-Bernstein and (p, q)-Bernstein operators and the Kelisky-Rivlin type theorem. Filomat **32**(12), 4351–4364 (2018)
36. I.A. Rus, Iterates of Stancu operators, via contraction principle. Stud. Univ. Babes-Bolyai Math. **47**, 101–104 (2002)
37. I.A. Rus, Iterates of Bernstein operators, via contraction principle. J. Math. Anal. Appl. **292**, 259–261 (2004)
38. A. Sultana, V. Vetrivel, Fixed points of Mizoguchi-Takahashi contraction on a metric space with a graph and applications. J. Math. Anal. Appl. **417**, 336–344 (2014)
39. Y. Wang, Y. Zhou, Iterates properties for q-Bernstein-Stancu operators. Int. J. Model. Optim. **3**(4), 362 (2013)
40. J. Yadav, S.A. Mohiuddine, A. Kajla, A. Alotaibi, $\alpha$-Bernstein-integral type operators. Bull. Iran. Math. Soc. **49**(5), 59 (2023)

# Existence of Solutions for Infinite System of Nonlinear $q$-Fractional Boundary Value Problem via Generalized Darbo's Fixed Point Theorem

**Anupam Das and Sudip Deb**

**Abstract** In this chapter, we use a contraction operator to prove a generalized Darbo-type fixed point theorem with the intention to study the existence of solutions of an infinite system of nonlinear $q$-fractional boundary value problem in Banach spaces formulated with the help of measure of noncompactness. Moreover, a suitable example is presented in order to illustrate the existence result.

**Keywords** Sequence space · Darbo-type theorem · Banach space · Measure of noncompactness

**Mathematics Subject Classification (2010)** 47H10 · 45G15 · 46B45 · 47H08

## 1 Introduction and Preliminaries

Kuratowski [14] was the first who described the idea of measure of noncompactness (MNC). Darbo [11] presented the world with a fixed point theorem making use of the concepts of MNC, later popularly known by his name as the "Darbo fixed point theorem". Since then, it has become a crucial tool for researchers while studying existence or solvability of nonlinear functional equations forming new Banach Spaces and employing "Darbo fixed point theorem" on them. Recently, Banaś and Lecko [9], Mohiuddine et al. [15] and Mursaleen et al. [16, 17] made efforts on studying infinite system of special kinds of differential equations to obtain the existence of solutions in Banach spaces like in the spaces of null $c_0$, convergent $c$, absolutely $p$-summable $\ell_p$ and $\ell_1$ (the case where $p = 1$).

---

A. Das (✉) · S. Deb
Department of Mathematics, Cotton University, Guwahati, Assam, India
e-mail: math.anupam@gmail.com

S. Deb
e-mail: math.sudip2022@gmail.com; mth2191003_sudip@cottonuniversity.ac.in



S. A. Mohiuddine et al. (eds.), *Positive Operators and Fixed-Point Theorems with Applications*, Industrial and Applied Mathematics,
https://doi.org/10.1007/978-981-96-2476-8_3

For studying real-world problems in science and engineering such as the theory of neural nets, kinetic theory of gas, radiation, mechanics, and neutron transportation integral and differential equations are of great help. Most recently, using infinite system integral and differential equations along with MNCs on specified Banach Spaces, many researches have shown existence of solutions to various real-world problems [6, 12, 18].

Let $\mathfrak{E}$ be a Banach space and $\mathbb{B}(\theta, \hat{r}) = \{x \in \mathfrak{E} : \|x - \theta\| \leq \hat{r}\}$ be a closed ball in $\mathfrak{E}$. If $\mathcal{Y} (\neq \phi) \subseteq \mathfrak{E}$, $\bar{\mathcal{Y}}$ and Conv$\mathcal{Y}$ denote the closure and the convex closure of the set $\mathcal{Y}$. Further, the family of nonempty bounded and relatively compact spaces are expressed with the symbols $\mathfrak{M}_{\mathfrak{E}}$ and $\mathfrak{N}_{\mathfrak{E}}$ respectively.

The definition of MNC in [7] (see also [10]) says the following.

**Definition 1.1** A mapping $\mathcal{Q} : \mathfrak{M}_{\mathfrak{E}} \to \mathbb{R}_+ (= [0, \infty))$ is called MNC in $\mathfrak{E}$ if

(i) $\mathcal{J} \in \mathfrak{M}_{\mathfrak{E}}$ implies $\mathcal{Q}(\mathcal{J}) = 0$ gives $\mathcal{J}$ be relatively compact.
(ii) $\ker \mathcal{Q} = \{\mathcal{J} \in \mathfrak{M}_{\mathfrak{E}} : \mathcal{Q}(\mathcal{J}) = 0\} \neq \phi$. Also $\ker \mathcal{Q} \subset \mathfrak{N}_{\mathfrak{E}}$.
(iii) $\mathcal{J} \subseteq \mathcal{J}_1 \implies \mathcal{Q}(\mathcal{J}) \leq \mathcal{Q}(\mathcal{J}_1)$.
(iv) $\mathcal{Q}(\bar{\mathcal{J}}) = \mathcal{Q}(\mathcal{J})$.
(v) $\mathcal{Q}(Conv\mathcal{J}) = \mathcal{Q}(\mathcal{J})$.
(vi) $\mathcal{Q}(\varsigma\mathcal{J} + (1-\varsigma)\mathcal{J}_1) \leq \varsigma\mathcal{Q}(\mathcal{J}) + (1-\varsigma)\mathcal{Q}(\mathcal{J}_1)$, $\varsigma \in [0, 1]$.
(vii) For a nested sequence of sets $\mathcal{S}_k \in \mathfrak{M}_{\mathfrak{E}}$, where $\mathcal{S}_{k+1} \subset \mathcal{S}_k$ for $k \in \mathbb{N}$ and $\mathcal{S}_k = \bar{\mathcal{S}}_k$ such that $\lim_{k\to\infty} \mathcal{Q}(\mathcal{S}_k) = 0$ then $\bigcap_{k=1}^{\infty} \mathcal{S}_k \neq \phi$.

Now we recall a Banach space with it's respective norm, namely

$$l_1 = \left\{ \{x_k\} : \sum_{k=1}^{\infty} |x_k| < \infty \right\},$$

i.e. $l_1$ is the collection of all sequences $\{x_K\}$ of scalars for which $\sum_{k=1}^{\infty} |x_k|$ is finite. Hence, $l_1$ is a Banach space with the norm

$$\| \{x_k\} \| = \sum_{k=1}^{\infty} |x_k|.$$

In [7] the measure of noncompactness $\chi_{l_1}$ in $(l_1, \| \|)$ is given by

$$\chi_{l_1}(B) = \lim_{n\to\infty} \left[ \sup_{y\in B} \left( \sum_{k\geq n} |y_k| \right) \right]. \tag{1.1}$$

We will use $C(I, l_1)$ to denote the collection of all continuous mappings from $I = [0, 1]$ to $l_1$ and $C(I, l_1)$ is a Banach space with the norm

$$\|x\|_{C(I,l_1)} = \sup_{s\in I} \|x(s)\|_{l_1},$$

where $x(s) = (x_n(s))_{n=1}^{\infty} \in C(I, l_1)$. For any nonempty bounded set $E$ of $C(I, l_1)$ and for $s \in I$, one defines $E(s) = \{x(s) : x(s) \in E\}$ and hence its MNC is given by

$$\chi_{C(I,l_1)}(E) = \sup_{s \in I} \chi_{l_1}(E(s)).$$

Recall the theorem given in [11] as follows.

**Theorem 1.1** *For a nonempty, closed, bounded and convex subset $\mathcal{J}$ of $\mathbb{E}$, let $\mathfrak{S} : \mathcal{J} \to \mathcal{J}$ be a continuous mapping and for some $\kappa \in [0, 1)$ we have*

$$\mathcal{Q}(\mathfrak{S}\Lambda) \leq \kappa \mathcal{Q}(\Lambda), \ \ \Lambda \subseteq \mathcal{J}.$$

*Then $\mathfrak{S}$ has a fixed point.*

## 2 Dorbo-Type Fixed Point Theorems

We start this section recalling certain class of functions quite essential for proving our generalized form of Dorbo fixed theorem. This has already been discussed in [13].

**Definition 2.1** Let $\hat{F}$be the collection of all functions $\bar{O} : \mathbb{R}_+ \times \mathbb{R}_+ \to \mathbb{R}_+$ where $\mathbb{R}_+ = [0, \infty)$ such that

(1) $Max\{a, b\} \leq \bar{O}(a, b), \ \ a, b \geq 0.$
(2) $\bar{O}$ is continuous and non-decreasing.
(3) $\bar{O}(a_1 + a_2, b_1 + b_2) \leq \bar{O}(a_1, b_1) + \bar{O}(a_2, b_2).$

For example, $\bar{O}(a, b) = a + b$.

**Theorem 2.1** *For a non-empty, closed, bounded convex subset $U$ of a Banach space $\mathbb{E}$ with an arbitrary MNC $\mathcal{G}$, let $\mathcal{T} : U \to U$ be a continuous mapping satisfying the inequality*

$$\psi\left[\bar{O}(\mathcal{Q}(\mathcal{T}X), \gamma(\mathcal{Q}(\mathcal{T}X)))\right] \leq \phi\left(\bar{O}(\mathcal{Q}(X), \gamma(\mathcal{Q}(X)))\right) - \theta\left(\bar{O}(\mathcal{Q}(X), \gamma(\mathcal{Q}(X)))\right) \quad (2.1)$$

*for any $X (\neq \phi) \subseteq U$, $\bar{O} \in \hat{F}$, $\mathcal{G}$ being any arbitrary MNC and $\phi, \theta, \gamma, \psi : \mathbb{R}_+ \to \mathbb{R}_+$ such that $\phi, \theta$ are bounded on any boundary interval of $\mathbb{R}_+$ and $\psi, \gamma$ being continuous. Moreover, assume that*

*(1) $\psi(x) \leq \phi(y) \Rightarrow x \leq y$.*
*(2) For any sequence $x_n$ in $\mathbb{R}_+$ with $x_n \to t > 0$, $\psi(t) - \lim_{n\to\infty} \phi(x_n) + \lim_{n\to\infty} \theta(x_n) > 0$.*

*Then*

$$\mathcal{T} \text{ has at least one fixed point in } U. \quad (2.2)$$

***Proof*** We form a sequence of sets $\{U_n\}_{n=1}^{\infty}$ such that $U_1 = D$ and $U_{n+1} = Conv(\mathcal{T}U_n)$ for $n \in \mathbb{N}$. Clearly we have $\mathbb{T}D_1 = \mathbb{T}D \subseteq U = U_1$, $U_2 = Conv(\mathcal{T}U_1) \subseteq U = U_1$. Proceeding this way we get the following nested sequence of sets,

$$U_1 \supseteq u_2 \supseteq U_3 \supseteq \ldots \supseteq U_n \supseteq U_{n+1} \supseteq \ldots.$$

Suppose $\exists\ n_0 \in \mathbb{N}$ with $\mathcal{Q}(U_{n_0}) = 0$, then $U_{n_0}$ is compact. And using Schauder theorem [5], we can tell that $\mathcal{T}$ has a fixed point in $U \subseteq \mathbb{E}$.

Otherwise, $\mathcal{Q}(U_n) > 0$ for $n \in \mathbb{N}$. The sequence $\{\mathcal{Q}(U_n)\}_{n=1}^{\infty}$ is nonnegative, decreasing and bounded below. Hence, $\{\mathcal{Q}(U_n)\}_{n=1}^{\infty}$ is convergent. Inequality (2.1) gives

$$\begin{aligned}\psi\left[\bar{O}\{\mathcal{Q}(U_{n+1}), \gamma(\mathcal{Q}(U_{n+1}))\}\right] &= \psi\left[\bar{O}\{\mathcal{Q}(Conv\mathcal{T}U_n), \gamma(\mathcal{Q}(Conv\mathcal{T}U_n))\}\right] \\ &= \psi\left[\bar{O}\{\mathcal{Q}(\mathcal{T}U_n), \gamma(\mathcal{Q}(\mathcal{T}U_n))\}\right] \\ &\leq \phi\left[\bar{O}\{\mathcal{Q}(U_n), \gamma(\mathcal{Q}(U_n))\}\right] - \theta\left[\bar{O}\{\mathcal{Q}(U_n), \gamma(\mathcal{Q}(U_n))\}\right].\end{aligned} \tag{2.3}$$

As $\theta \geq 0$, thus from 2.3 we get

$$\psi\left[\bar{O}\{\mathcal{Q}(U_{n+1}), \gamma(\mathcal{Q}(U_{n+1}))\}\right] \leq \phi\left[\bar{O}\{\mathcal{Q}(U_n), \gamma(\mathcal{Q}(U_n))\}\right]$$

Now, from the condition (1), we get

$$\bar{O}\{\mathcal{Q}(U_{n+1}), \gamma(\mathcal{Q}(U_{n+1}))\} \leq \bar{O}\{\mathcal{Q}(U_n), \gamma(\mathcal{Q}(U_n))\}.$$

Hence, $\{\bar{O}(\mathcal{Q}(U_n), \gamma(\mathcal{Q}(U_n)))\}_{n=1}^{\infty}$ is a decreasing sequence of positive real numbers. So $\exists\ r \geq 0$ such that

$$\lim_{n\to\infty} \bar{O}(\mathcal{Q}(U_n), \gamma(\mathcal{Q}(U_n))) = r. \tag{2.4}$$

By taking $n \to \infty$ on 2.3 we get,

$$\psi(r) \leq \lim_{n\to\infty} \phi(r_n) + \lim_{n\to\infty} \theta(r_n)$$

where $r_n = \bar{O}\{\mathcal{Q}(U_n), \gamma(\mathcal{Q}(U_n))\}$,
which is a contradiction of condition (2) if $r > 0$.
Hence, r = 0.
That is,

$$\begin{aligned}&\lim_{n\to\infty} \bar{O}(\mathcal{Q}(U_n), \gamma(\mathcal{Q}(U_n))) = 0 \quad [By\ using\ 2.4] \\ &\Rightarrow \lim_{n\to\infty} \mathcal{Q}(u_n) = 0.\end{aligned}$$

Now since $U_{n+1} \subseteq U_n and \mathcal{T}U_n \subseteq U_n$, from condition (vii) of the definition (2.1), we conclude that $U_\infty = \bigcap_{n=1}^{\infty} U_n$ is non-empty, closed, convex and invariant under the operator $\mathcal{T}$ and belongs to ker$\mathcal{Q}$. So, we can say that $\mathcal{T}$ has a fixed point in U. □

**Theorem 2.2** *For a non-empty, closed, bounded convex set $U$ of a Banach space $\mathbb{E}$ with an arbitrary MNC $\mathcal{Q}$, if $\mathcal{T}: U \to U$ is a continuous mapping having the inequality*

$$\psi\left[\mathcal{Q}(\mathcal{T}X)+\gamma(\mathcal{Q}(\mathcal{T}X))\right] \leq \phi\left[\mathcal{Q}(X)+\gamma(\mathcal{Q}(X))\right]-\theta\left[\mathcal{Q}(X)+\gamma(\mathcal{Q}(X))\right] \quad (2.5)$$

*for any $X(\neq\phi) \subseteq U$, $\mathcal{G}$ is any arbitrary MNC and $\phi, \theta, \gamma, \psi : \mathbb{R}_+ \to \mathbb{R}_+$ such that $\phi, \theta$ are bounded on any boundary interval of $\mathbb{R}_+$ and $\psi, \gamma$ are continuous. Moreover, assume that*

*(1)* $\psi(x) \leq \phi(y) \Rightarrow x \leq y$.
*(2) For any sequence $x_n$ in $\mathbb{R}_+$ with $x_n \to t > 0$ ,* $\psi(t) - \lim_{n\to\infty} \phi(x_n) + \lim_{n\to\infty} \theta(x_n) > 0$.

*Then*

$$\mathcal{T} \text{ has at least one fixed point in } U. \quad (2.6)$$

***Proof*** We can prove this result considering the function $\bar{O}\,(a, b) = a + b$ in Theorem 2.1. □

**Theorem 2.3** *For a non-empty, closed, bounded convex set $U$ of a Banach space $\mathbb{E}$ with an arbitrary MNC $\mathcal{Q}$, if $\mathcal{T}: U \to U$ is a continuous mapping having the inequality*

$$\psi\,(\mathcal{Q}(\mathcal{T}X)) \leq \phi\,(\mathcal{Q}(X)) - \theta\,(\mathcal{Q}(X)) \quad (2.7)$$

*for any $X(\neq\phi) \subseteq U$, $\mathcal{G}$ is any arbitrary MNC and $\phi, \theta, \psi : \mathbb{R}_+ \to \mathbb{R}_+$ such that $\phi, \theta$ are bounded on any boundary interval of $\mathbb{R}_+$ and $\psi$ are continuous. Moreover, assume that*

*(1)* $\psi(x) \leq \phi(y) \Rightarrow x \leq y$.
*(2) For any sequence $x_n$ in $\mathbb{R}_+$ with $x_n \to t > 0$ ,* $\psi(t) - \lim_{n\to\infty} \phi(x_n) + \lim_{n\to\infty} \theta(x_n) > 0$.

*Then*

$$\mathcal{T} \text{ has at least one fixed point in } U. \quad (2.8)$$

***Proof*** Taking the function $\gamma(s) = 0, s \in \mathbb{R}_+$ in Theorem 2.2, we get the required result. □

**Remark 2.1** Taking $\psi(s) = s, \phi(s) = s, s \in \mathbb{R}_+$ and $\theta(t) = (1-k)t, k \in [0, 1)$ in theorem 2.3 we obtain

$$\mathcal{Q}(\mathcal{T}X) \leq \mathcal{Q}(X) - (1-k)\mathcal{Q}(X),$$

i.e.

$$\mathcal{Q}(\mathcal{T}X) \leq k\mathcal{Q}(X), k \in [0, 1).$$

Hence, we obtain Darbo's theorem.

## 3 Preliminaries

In this section, we recall some definitions and lemmas from fractional $q$-calculus. For greater understanding, the reader might read these literature References [1–4].

Let $q \in (0, 1)$ and define

$$[\alpha]_q = \frac{1 - q^\alpha}{1 - q}, \quad a \in \mathbb{R}.$$

The $q$-analog of the power function $(a - b)^n$ with $n = 0, 1, 2, \ldots$ is

$$(a - b)^0 = 1, \ (a - b)^n = \prod_{k=0}^{n-1} (a - bq^k), \ n \in \mathbb{N}, \ a, b \in \mathbb{R}.$$

More generally, if $\alpha \in \mathbb{R}$, then

$$(a - b)^{(\alpha)} = a^\alpha \prod_{n=0}^{\infty} \frac{a - bq^n}{a - bq^{\alpha+n}}.$$

Note that if $b = 0$ then $a^{(\alpha)} = a^\alpha$. The $q$-gamma function is defined by

$$\Gamma_q(x) = \frac{(1 - q)^{(x-1)}}{(1 - q)^{x-1}}, \quad x \in \mathbb{R} \setminus \{0, -1, -2, \ldots\},$$

and satisfies $\Gamma_q(x + 1) = [x]_q \, \Gamma_q(x)$.

The $q$-derivative of a function $f$ is defined by

$$(D_q f)(x) = \frac{f(x) - f(qx)}{(1 - q)x}$$

and the $q$-integral of a function $f$ defined in the interval $[0, b]$ is given by

$$(I_q f)(x) = \int_0^x f(t) d_q t = x(1 - q) \sum_{n=0}^{\infty} f(xq^n) q^n, \ x \in [0, b].$$

**Lemma 3.1** *If $f : [0, 1] \to \mathbb{R}$ is a continuous function, then*

$$\left| \int_0^x f(t) d_q t \right| \leq \int_0^x |f(t)| \, d_q t, \; x \in [0, 1].$$

**Lemma 3.2** *If $\beta > 0$ and $0 \leq a \leq b \leq t$ then*

$$(t - b)^{(\beta)} \leq (t - a)^{(\beta)}.$$

**Definition 3.1** For $\alpha \geq 0$ and $f : [0, 1] \to \mathbb{R}$, the fractional $q$-integral of Riemann-Liouville type is defined as

$$(I_q^0 f)(t) = f(t)$$

and

$$(I_q^\alpha f)(t) = \frac{1}{\Gamma_q(\alpha)} \int_0^t (t - qs)^{(\alpha - 1)} f(s) d_q s, \; \alpha > 0.$$

**Definition 3.2** The fractional $q$-derivative of Riemann-Liouville type of order $\alpha \geq 0$ is defined by

$$(D_q^\alpha f)(t) = \left( D_q^{[\alpha]} I_q^{[\alpha] - \alpha} f \right)(t), \; \alpha > 0, \; t \in [0, 1]$$

where $[\alpha]$ is the smallest integer greater than or equal to $\alpha$. Evidently, $(D_q^\alpha f)(t) = (D_q f)(t)$ when $\alpha = 1$.

For a nonnegative continuous function $f : [0, 1] \times \mathbb{R} \to \mathbb{R}$, we consider the following nonlinear $q$-fractional boundary value problem [4],

$$(D_q^\alpha y)(x) = -f(x, y(x)), \; 0 < x < 1, \; y(0) = y(1) = 0 \tag{3.1}$$

where $1 < \alpha \leq 2, \; 0 < q < 1$.

$y$ is a solution of the boundary value problem (3.1) if and only if $y$ satisfies the following integral equation:

$$y(x) = \int_0^1 G(x, qt) f(t, y(t)) d_q t$$

where the function $G$ is defined by

$$G(x, t) = \frac{1}{\Gamma_q(\alpha)} \left\{ (x(1 - t))^{(\alpha - 1)} - (x - t)^{(\alpha - 1)} \right\}, \; 0 \leq t \leq x \leq 1$$

and

$$G(x, t) = \frac{1}{\Gamma_q(\alpha)} (x(1 - t))^{(\alpha - 1)}, \; 0 \leq x \leq t \leq 1.$$

Here, we are investigating the existence of solutions to an infinite system of $q$-fractional differential equations with boundary conditions which is considered as follows:

$$D_q^{\alpha} y_n(x) = -f_n(x, y(x)),\ 0 < x < 1,\ y_n(0) = y_n(1) = 0 \tag{3.2}$$

where $1 < \alpha \leq 2,\ 0 < q < 1$ and $f_n : [0, 1] \times E \to \mathbb{R}$ are nonnegative continuous functions for all $n \in \mathbb{N}$. Also $y(x) = (y_n(x))_{n=1}^{\infty} \in E$ where $E$ is a sequence space.

$y_n(x)$ is a solution of the boundary value problem (3.2) if and only if $y_n(x)$ satisfies the following:

$$y_n(x) = \int_0^1 G(x, qt) f_n(t, y(t)) d_q t. \tag{3.3}$$

## 4 Existence of Solutions for an Infinite System of $q$-Fractional Differential Equations with Boundary Conditions

In order to prove our desired result, we need to consider the following assumptions:

(1) $\forall n \in \mathbb{N},\ f_n : I \times C(I, l_1) \to \mathbb{R}_+ = [0, \infty)$ are continuous where $I = [0, 1]$. The operator $f$ is defined from $I \times C(I, l_1)$ to $C(I, l_1)$ as

$$(x, y(x)) \to (fy)(x) = (f_n(x, y(x)))_{n=1}^{\infty}$$

which is the class of all functions $((fy)(x))_{x \in I}$ equicontinuous at every point of the space $C(I, l_1)$.

(2) For every $y(x) \in C(I, l_1),\ n \in \mathbb{N},\ x \in I$, we have

$$|f_n| \leq A_n(x) + B_n(x)\, |y_n(x)|$$

where both $A_n(x),\ B_n(x)$ are nonnegative real continuous functions defined on $I$ such that the sequence $\sum_{n=1}^{\infty} A_n(x)$ is uniformly convergent on $I$.

Further we assume that

$$\bar{A}(x) = \sum_{n=1}^{\infty} A_n,\ \bar{A} = \sup_{x \in I} \bar{A}(x),\ \bar{B} = \sup_{n \in \mathbb{N},\ x \in I} B_n(x)\ with\ \bar{B} < \Gamma_q(\alpha).$$

**Theorem 4.1** *Under assumptions (1)–(2), the infinite system (3.2) admits a solution in* $C(I, l_1)$.

***Proof*** We have $(x(1-t))^{(\alpha-1)} \leq (x)^{(\alpha-1)} = x^{\alpha-1} \leq 1$.

Therefore

$$(x(1-t))^{(\alpha-1)} - (x-t)^{(\alpha-1)} \leq x^{\alpha-1} - (x-t)^{(\alpha-1)} \leq x^{\alpha-1} \leq 1$$

and

$$(x(1-t))^{(\alpha-1)} \leq x^{\alpha-1} \leq 1.$$

Thus we have

$$|G(x,t)| \leq \frac{1}{\Gamma_q(\alpha)}.$$

For arbitrary fixed $x \in I$, using (2) and (3.2)

$$\begin{aligned}
\|y(x)\|_{l_1} &= \sum_{n\geq 1}\left|\int_0^1 G(x,qt) f_n(t,y(t))d_qt\right| \\
&\leq \sum_{n\geq 1}\int_0^1 |G(x,qt)|\, f_n(t,y(t))d_qt \\
&\leq \frac{1}{\Gamma_q(\alpha)}\sum_{n\geq 1}\int_0^1 f_n(t,y(t))d_qt \\
&\leq \frac{1}{\Gamma_q(\alpha)}\sum_{n\geq 1}\int_0^1 \{A_n(t) + B_n(t)\,|y_n(t)|\}\, d_qt \\
&\leq \frac{1}{\Gamma_q(\alpha)}\int_0^1 \left\{\bar{A} + \bar{B} \parallel y \parallel_{C(I,l_1)}\right\} d_qt \\
&= \frac{\bar{A} + \bar{B} \parallel y \parallel_{C(I,l_1)}}{\Gamma_q(\alpha)}
\end{aligned}$$

which gives

$$(\Gamma_q(\alpha) - \bar{B}) \parallel y \parallel_{C(I,l_1)} \leq \bar{A}$$

,

i.e.

$$\|y\|_{C(I,l_1)} \leq \frac{\bar{A}}{\Gamma_q(\alpha) - \bar{B}} = d(say).$$

Suppose

$$B = \left\{y \in C\,(I,l_1) : \|y\|_{C(I,l_1)} \leq d\right\}$$

which is a nonempty, closed, bounded and convex subset of $C\,(I,l_1)$.

For any fixed $x \in I$, we define the operator from $C(I, l_1)$ to $C(I, l_1)$ as follows:

$$(Ty)(x) = \{(T_n y)(x)\}_{n=1}^{\infty} = \left\{ \int_0^1 G(x, qt) f_n(t, y(t)) d_q t \right\}_{n=1}^{\infty}$$

where $y(x) = \{y_n(x)\}_{n=1}^{\infty} \in C(I, l_1)$ and $y_n(x) \in C(I, \mathbb{R})$.

As $\{f_n(x, y(x))\}_{n=1}^{\infty} \in C(I, l_1)$ we have

$$\begin{aligned}
\sum_{n \geq 1} \left|(T_n y)(x)\right| &= \sum_{n \geq 1} \left| \int_0^1 G(x, qt) f_n(t, y(t)) d_q t \right| \\
&\leq \frac{1}{\Gamma_q(\alpha)} \sum_{n \geq 1} \int_0^1 \{A_n(t) + B_n(t) \left|y_n(t)\right|\} d_q t \\
&\leq \frac{1}{\Gamma_q(\alpha)} \int_0^1 \left\{\bar{A} + \bar{B} \parallel y \parallel_{C(I, l_1)}\right\} d_q t \\
&= \frac{\bar{A} + \bar{B} d}{\Gamma_q(\alpha)}.
\end{aligned}$$

Hence $(Ty)(x) \in C(I, l_1)$.

Also $(T_n y)(x)$ satisfies the boundary conditions

$$(T_n y)(0) = \int_0^1 G(0, qt) f_n(t, y(t)) d_q t = \int_0^1 0.f_n(t, y(t)) d_q t = 0$$

and

$$(T_n y)(1) = \int_0^1 G(1, qt) f_n(t, y(t)) d_q t = \int_0^1 0.f_n(t, y(t)) d_q t = 0.$$

Again, $T$ maps $B$ to $B$ itself.

Taking the assumption (1) in our consideration, we take $z(x) = \{z_n(x)\}_{n=1}^{\infty} \in B$ and $\exists\, \epsilon > 0$ for each $\delta > 0$ satisfying

$$\parallel (fy)(x) - (fz)(x) \parallel_{l_1} \leq \epsilon \Gamma_q(\alpha) \text{ for } y(x),\ z(x) \in B$$

whenever $\parallel y(x) - z(x) \parallel_{l_1} \leq \delta$ for all $x \in I$.

For any fixed $x \in I$,

$$\parallel (Ty)(x) - (Tz)(x) \parallel_{l_1} \leq \sum_{n \geq 1} \int_0^1 \left|G(x, qt)\right| \left|f_n(t, y(t)) - f_n(t, y(t))\right| d_q t < \epsilon$$

thus $T$ is continuous in $B$ as $x$ is arbitrarily fixed.

Now,

$$\chi_{l_1}(TB) = \lim_{n \to \infty} \sup_{y \in B} \sum_{k \geq n} \left| \int_0^1 G(x, qt) f_n(t, y(t) d_q) t \right|$$

$$
\begin{aligned}
&\leq \frac{1}{\Gamma_q(\alpha)} \lim_{n\to\infty} \sup_{y\in B} \sum_{k\geq n} \int_0^1 \left\{A_n(t) + B_n(t)\,|y_n(t)|\, d_q t\right\} \\
&\leq \frac{\bar{B}}{\Gamma_q(\alpha)} \lim_{n\to\infty} \sup_{y\in B} \sum_{k\geq n} \int_0^1 |y_n(t)|\, d_q t \\
&\leq \frac{\bar{B}}{\Gamma_q(\alpha)} \chi_{l_1}(B),
\end{aligned}
$$

which gives

$$
\sup_{x\in I} \chi_{l_1}(TB) \leq \sup_{x\in I} \frac{\bar{B}}{\Gamma_q(\alpha)} \chi_{l_1}(B), \tag{4.1}
$$

and implies

$$
\chi_{C(I,l_1)}(TB) \leq \frac{\bar{B}}{\Gamma_q(\alpha)} \chi_{C(I,c_1)}(B). \tag{4.2}
$$

Now using assumption (2) and Remark 2.1 we can say $\exists$ is a fixed point for $T$ in $B \subseteq C(I, l_1)$, i.e. the system we considered has a solution in $C(I, l_1)$. □

**Example 4.1** To validate the last Theorem 4.1, we suppose an infinite system of $q$-fractional boundary value problem as

$$
D_{\frac{1}{2}}^{\frac{3}{2}} y_n(x) = -\frac{x}{n^3} - \frac{1}{3}\Gamma_{\frac{1}{2}}\left(\frac{3}{2}\right)\left(\frac{1}{n^2} + 1\right) y_n(x),\ y_n(0) = y_n(1) = 0, \tag{4.3}
$$

for $x \in [0, 1] = I$ and $n \in \mathbb{N}$.

And we write

$$
f_n(x, y(x)) = -\frac{x}{n^3} - \frac{1}{3}\Gamma_{\frac{1}{2}}\left(\frac{3}{2}\right)\left(\frac{1}{n^2} + 1\right) y_n(x)
$$

and

$$
q = \frac{1}{2},\ \alpha = \frac{3}{2}.
$$

Here

$$
|f_n(x, y(x))| \leq \frac{1}{n^3} + \frac{1}{3}\Gamma_{\frac{1}{2}}\left(\frac{3}{2}\right)\left(\frac{1}{n^2} + 1\right) |y_n(x)|
$$

which implies

$A_n(x) = \frac{1}{n^3}$, $B_n(x) = \frac{1}{3}\Gamma_{\frac{1}{2}}\left(\frac{3}{2}\right)\left(\frac{1}{n^2}+1\right)$. Clearly the sequence $\sum_{n=1}^{\infty}\frac{1}{n^3}$ is uniformly convergent on $I = [0, 1]$. Also,

$$\bar{A} = 1, \ \bar{B} = \frac{2}{3}\Gamma_{\frac{1}{2}}\left(\frac{3}{2}\right)$$

and $\bar{B} < \Gamma_{\frac{1}{2}}\left(\frac{3}{2}\right)$.

If $y(x) \in C(I, l_1)$ then for all $x \in I$, we have

$$\sum_{n\geq 1} |f_n(x, y(x))| \leq \sum_{n\geq 1} \frac{1}{n^3} + \frac{2}{3}\Gamma_{\frac{1}{2}}\left(\frac{3}{2}\right) \sum_{n\geq 1} |y_n(x)|.$$

Since both the series on the right-hand side of the above expression are convergent, therefore $f_n(x, y(x)) \in C(I, l_1)$.

Let $z(x) = \{z_n(x)\}_{n=1}^{\infty} \in C(I, l_1)$ and $\epsilon > 0$.

For arbitrary fixed $x \in I$ we have

$$\| (fy)(x) - (fz)(x) \|_{l_1} \leq \sum_{n\geq 1} |f_n(x, y(x)) - f_n(x, z(x))| \leq \frac{2}{3}\Gamma_{\frac{1}{2}}\left(\frac{3}{2}\right) \| y(x) - z(x) \|_{l_1}.$$

Thus

$$\| (fy)(x) - (fz)(x) \|_{C(I,l_1)} < \epsilon$$

whenever

$$\| y - z \|_{C(I,l_1)} < \frac{3\epsilon}{2\Gamma_{\frac{1}{2}}\left(\frac{3}{2}\right)}$$

which proves the equicontinuity of $((fy)(x))_{x\in I}$ on $C(I, l_1)$. Since all the required conditions are satisfied, using Theorem 4.1 we can conclude that the considered system (4.3) has a solution in $C(I, l_1)$.

## References


1. R.P. Agarwal, Certain fractional q-integrals and q-derivatives. Proc. Camb. Philos. Soc. **66**, 365–370 (1969)
2. W.A. Al-Salam, Some fractional q-integrals and q-derivatives. Proc. Edinb. Math. Soc. **15**, 135–140 (1966)
3. M.H. Annaby, Z.S. Mansour, *q-Fractional Calculus and Equations*. Lecture Notes in Mathematics, vol. 2056 (Springer, Berlin, 2012)
4. R.A.C. Ferreira, Nontrivial solutions for fractional $q$-difference boundary value problem. Electron J quant Theory Diff. Equ. **70**, 1–10 (2010)
5. R.P. Agarwal, D. O'Regan, *Fixed Point Theory and Applications* (Cambridge University Press, 2004)
6. R. Arab, H.K. Nashine, N.H. Can, T.T. Binh, Solvability of functional-integral equations (fractional order) using measure of noncompactness. Adv. Diff. Equ. **2020**, 12 (2020)

7. J. Banaś, K. Goebel, *Measure of Noncompactness in Banach Spaces*. Lecture Notes in Pure and Applied Mathematics, vol. 60. (Marcel Dekker, New York, 1980)
8. J. Banaś, M. Krajewska, Existence of solutions for infinite systems of differential equations in spaces of tempered sequences. Electron. J. Diff. Equ. **60**, 1–28 (2017)
9. J. Banaś, M. Lecko, Solvability of infinite systems of differential equations in Banach sequence spaces. J. Comput. Appl. Math. **137**, 363–375 (2001)
10. J. Banaś, M. Mursaleen, Sequence spaces and measures of noncompactness with applications to differential and integral equations (Springer, 2014)
11. G. Darbo, Punti uniti in trasformazioni a codominio non compatto (Italian). Rend. Sem. Mat. Univ. Padova **24**, 84–92 (1955)
12. B. Hazarika, R. Arab, M. Mursaleen, Applications of measure of noncompactness and operator type contraction for existence of solution of functional integral equations. Complex Anal. Oper. Theory **13**, 3837–3851 (2019)
13. I. Altun, D. Turkoglu, A fixed point theorem for mappings satisfying a general contractive condition of operator type. J. comput. Anal. Appl. **9**(1), 9–14 (2007)
14. K. Kuratowski, Sur les espaces complets. Fund. Math. **15**, 301–309 (1930)
15. S.A. Mohiuddine, H.M. Srivastava, A. Alotaibi, Application of measures of noncompactness to the infinite system of second-order differential equations in $\ell_p$ spaces. Adv. Diff. Equ. **2016**, 317 (2016)
16. M. Mursaleen, S.M.H. Rizvi, Solvability of infinite system of second order differential equations in $c_0$ and $\ell_1$ by Meir-Keeler condensing operator. Proc. Am. Math. Soc. **144**(10), 4279–4289 (2016)
17. M. Mursaleen, S.M.H. Rizvi, B. Samet, Solvability of a class of boundary value problems in the space of convergent sequences. Appl. Anal. **97**, 1829–1845 (2018)
18. H.K. Nashine, R. Arab, R.P. Agarwal, A.S. Haghigh, Darbo type fixed and coupled fixed point results and its application to integral equation. Period. Math. Hungar. **77**, 94–107 (2018)

# On the Solutions of a Nonlinear Fractional Integral Equation Using the Darbo Fixed Point Theorem

**Bhuban Chandra Deuri and Anupam Das**

**Abstract** This paper generalizes a fixed point theorem using a newly designed contraction operator. In addition, this chapter describes the solvability of fractional integrals based on this generalized fixed point theorem along with appropriate examples.

**Keywords** Fractional integral equations · Measure of noncompactness(MNC) · Fixed point theorem

**MSC Subject Classification No** 45P05 · 45G10 · 47H10

## 1 Introduction

Kuratowski [18] was the first to suggest the notion of measure of noncompactness (MNC), which was essential to the development of the fixed point theory. Incorporating Kuratowski's MNC into his formulation, Darbo [3] expanded Schauder's fixed point theorem. Several authors then utilized these MNC in their research to investigate and determine numerous differential equations, integral equations, and integro-differential equations problems.

Banaś [1] axiomatic concept of the MNC, introduced in 1980, has sparked a lot of research interest in analyzing many problems.

Numerous researchers have generalized and utilized Darbo's theorem. This time, Darbo's theorem is generalized with an application. Numerous researchers and fractional order integral equations utilize MNC for the solution of multiple forms of integral equations(see [7, 13, 15, 17, 19, 20, 24]). Using C-class functions, Das

B. C. Deuri (✉)
Department of Mathematics, Rajiv Gandhi University, Doimukh, Arunachal Pradesh, India
e-mail: bhuban.deuri@rgu.ac.in

A. Das
Department of Mathematics, Cotton University, Guwahati, Assam, India
e-mail: math.anupam@gmail.com

 
S. A. Mohiuddine et al. (eds.), *Positive Operators and Fixed-Point Theorems with Applications*, Industrial and Applied Mathematics,
https://doi.org/10.1007/978-981-96-2476-8_4

et al. [21] developed several new generalizations. Parvaneh and Hussain [22] investigated the concept of generalized Wardowski type fixed point theorems using *alpha*-admissible FG-contractions in b-metric spaces. Recently, it has been established that integral equations are solvable (see [8–10, 14, 16, 26]).

In this paper, we give Darbo's fixed point theorem for solving fractional integral equations in Banach space using various functions. In this article, we also presented an appropriate illustration for validating the suggested theorem.

Let a real Banach space $(\mathscr{B}, \| \, . \, \|)$ and $B(c, r) = \{a \in \mathscr{B} : \| a - c \| \le r\}$. If $\mathscr{X}(\neq \emptyset) \subseteq \mathscr{B}$. Then

- $\bar{\mathscr{X}}$ = the closure of $\mathscr{X}$,
- $Conv\mathscr{X}$ = the convex closure of $\mathscr{X}$,
- $\mathfrak{M}_{\mathscr{B}}$ = collection of all non-empty and bounded subsets of $\mathscr{B}$,
- $\mathfrak{N}_{\mathscr{B}}$ = collection of all relatively compact sets,
- $\mathbb{R} = (-\infty, \infty)$,
  also
- $\mathbb{R}_+ = [0, \infty)$.

The definition of a MNC is as follows: [1].

**Definition 1.1** A function $\mu : \mathfrak{M}_{\mathscr{B}} \to [0, \infty)$ is said to be a MNC in $\mathscr{B}$ if it fulfills axioms:

(i) for all $\mathscr{X} \in \mathfrak{M}_{\mathscr{B}}$, we have $\mu(\mathscr{X}) = 0$ gives $\mathscr{X}$ is relatively compact.
(ii) $\ker \mu = \{\mathscr{X} \in \mathfrak{M}_{\mathscr{B}} : \mu(\mathscr{X}) = 0\} \neq \emptyset$ and $\ker \mu \subset \mathfrak{N}_{\mathscr{B}}$.
(iii) $\mathscr{X} \subseteq \mathscr{X}_1 \implies \mu(\mathscr{X}) \le \mu(\mathscr{X}_1)$.
(iv) $\mu(\bar{\mathscr{X}}) = \mu(\mathscr{X})$.
(v) $\mu(Conv\mathscr{X}) = \mu(\mathscr{X})$.
(vi) $\mu(\varpi\mathscr{X} + (1-\varpi)\mathscr{X}_1) \le \varpi\mu(\mathscr{X}) + (1-\varpi)\mu(\mathscr{X}_1)$ for $\varpi \in [0, 1]$.
(vii) if $\mathscr{X}_k \in \mathfrak{M}_{\mathscr{B}}$, $\mathscr{X}_k = \bar{\mathscr{X}_k}$, $\mathscr{X}_{k+1} \subset \mathscr{X}_k$ for $k = 1, 2, 3, 4, 5, \ldots$ and $\lim_{k\to\infty} \mu(\mathscr{X}_k) = 0$ then $\bigcap_{k=1}^{\infty} \mathscr{X}_k \neq \emptyset$.

The *kernel of measure* $\mu$ is defined to be the family $ker\,\mu$. Since $\mu(\mathscr{X}_\infty) \le \mu(\mathscr{X}_k)$, $\mu(\mathscr{X}_\infty) = 0$. So, $\mathscr{X}_\infty = \bigcap_{k=1}^{\infty} \mathscr{X}_k \in ker\,\mu$.

**Some important theorem and definition** The following are some fundamental theorems to recall:

**Theorem 1.1** (Shauder [2]) *Let $\mathscr{G}$ be a non-empty, bounded, closed and convex subset(NBCCS) of a Banach Space $\mathscr{B}$. If $\mathscr{T} : \mathscr{G} \to \mathscr{G}$ is a continuous and compact mapping, then it must have at least one fixed point.*

**Theorem 1.2** (Darbo [3]) *Let $\mathscr{G}$ be a NBCCS of a Banach Space $\mathscr{B}$. Assume $\mathscr{T} : \mathscr{G} \to \mathscr{G}$ is a continuous mapping with a constant $\varpi \in [0, 1)$ such that*

$$\mu(\mathscr{T}\mathscr{X}) \le \varpi\mu(\mathscr{X}), \mathscr{X} \subseteq \mathscr{G}.$$

*Then $\mathscr{T}$ has a fixed point.*

The following related concepts are needed to establish an extension of Darbo's fixed point theorem:

**Definition 1.2** ([4]) A function $\psi : \mathbb{R}_+ \to \mathbb{R}_+$ is an altering distance function if the subsequent properties hold true:
$(a)$ $\psi(t) = 0 \iff t = 0$,
$(b)$ $\psi$ is continuous and strictly increasing.
where, $\psi \in \Psi$ is the set of altering distance functions.

For example,
(1) $\psi(t) = (1 - k)t, 0 \le k < 1$.

**Definition 1.3** ([5]) A continuous function $\varphi : \mathbb{R}_+ \to \mathbb{R}_+$ is an ultra-altering distance function if $\varphi(0) \ge 0$, $\varphi(t) > 0$ and $t > 0$, where, $\varphi \in \Psi$.

**Definition 1.4** ([5]) A continuous function $\mathscr{F} : \mathbb{R}_+ \times \mathbb{R}_+ \to \mathbb{R}_+$ is a function of C-class, if subsequent axioms hold true:
$a$ $\mathscr{F}(p, w) \le p$;
$b$ $\mathscr{F}(p, w) = p$ implies that either $p = 0$ or $w = 0$, $p, w \in \mathbb{R}_+$.
Noticeably, $\mathscr{F}(0, 0) = 0$. A collection of C-class of functions is symbolized as C.

For example,
(1) $\mathscr{F}(p, w) = p - \frac{w}{k+w}, 0 \le k < 1$.
(2) $\mathscr{F}(p, w) = \frac{p-w}{1+w}$.

**Definition 1.5** ([5]) A continuous function $h : \mathbb{R}_+ \to \mathbb{R}_+$ is a function of C-class if $ht \ge t$, $t \in \mathbb{R}_+$.

For example,
(1) $h(t) = k^t - 1, k > 1$.
(2) $h(t) = mt, m \ge 1$.

## 2 Darbo Fixed Point Theory Results

**Theorem 2.1** *Suppose that $\mathscr{G}$ be a $\mathcal{NBCCS}$ of a Banach space $\mathscr{B}$ and, the function $\mathscr{T} : \mathscr{G} \to \mathscr{G}$ is a continuous such that*

$$h\{\psi(\mu(\mathscr{T}X))\} \le \mathscr{F}(\psi(\mu(X)), \varphi(\mu(X))) \tag{2.1}$$

*where $X \subset \mathscr{G}$ and $\mu$ is an arbitrary MNC and $h \in C, \psi \in \Psi, \varphi \in \Psi$ and $\mathscr{F} \in C$. So that $\mathscr{T}$ has atleast one fixed point in $\mathscr{G}$.*

***Proof*** Assume that a sequence $\{\mathscr{C}_k\}_{k=1}^{\infty}$ with $\mathscr{C}_0 = \mathscr{C}$ and $\mathscr{C}_k = Conv(\mathscr{T}\mathscr{C}_{k-1})$ for $k \ge 1$. We have, $\mathscr{T}\mathscr{C}_0 = \mathscr{T}\mathscr{C} \subseteq \mathscr{C}_0 = \mathscr{C}_0$, $\mathscr{C}_1 = Conv(\mathscr{T}\mathscr{C}_0) \subseteq \mathscr{C} = \mathscr{C}_0$; therefore, by proceeding in the same manner gives $\mathscr{C}_0 \supseteq \mathscr{C}_1 \supseteq \mathscr{C}_2 \supseteq \mathscr{C}_3 \supseteq \ldots \supseteq \mathscr{C}_k \supseteq \mathscr{C}_{k+1} \supseteq \ldots$.

If there exists $k_0 \in \mathbb{N}$ satisfying $\mathscr{F}\left(\psi(\mu(\mathscr{C}_{k_0})), \varphi(\mu(\mathscr{C}_{k_0}))\right) = 0$. Then $\mu(\mathscr{C}_{k_0}) = 0$, so $\mathscr{C}_{k_0}$ is a relatively compact set and $\mathscr{T}\mathscr{C}_{k_0} \subseteq Conv(\mathscr{T}\mathscr{C}_{k_0}) = \mathscr{C}_{k_0+1}$. Then, Theorem 1.1 (Schauder's) yeids that $\mathscr{T}$ has a fixed point in $\mathscr{C}$.

Again, if $\mathscr{F}\left(\mu(\mathscr{C}_k), \phi(\mu(\mathscr{C}_k))\right) > 0,\ k \in \mathbb{N}$.

By the Eq. 2.1, we now have

$$\begin{aligned} & h\left\{\psi\left(\mu(\mathscr{C}_{k+1})\right)\right\} \\ & = h\left\{\psi\left(\mu(Conv(\mathscr{T}\mathscr{C}_k))\right)\right\} \\ & = h\left\{\psi\left(\mu(\mathscr{T}\mathscr{C}_k)\right)\right\} \\ & \leq \mathscr{F}\left(\psi(\mu(\mathscr{C}_k)), \varphi(\mu(\mathscr{C}_k))\right) \\ & \leq \psi(\mu(\mathscr{C}_k)). \end{aligned}$$

Since $\psi$ is a continuous and strictly increasing mapping, we obtain

$$\mu(\mathscr{C}_{k+1}) \leq \mu(\mathscr{C}_k).$$

Then, $\{\mu(\mathscr{C}_k)\}$ is a bounded below and decreasing.
So, it is converges to $b = \inf\{\mathscr{C}_k\}$.

If possible let $b > 0$. By using (2.1), we get

$$\begin{aligned} h\left\{\psi\left(\mu(\mathscr{C}_{k+1})\right)\right\} &= h\left\{\psi\left(\mu(Conv(\mathscr{T}\mathscr{C}_k))\right)\right\} \\ &\leq \mathscr{F}\left(\psi(\mu(\mathscr{C}_k)), \varphi(\mu(\mathscr{C}_k))\right). \end{aligned}$$

As $k \to \infty$, we get

$$h\left\{\psi\left(b\right)\right\} \leq \mathscr{F}\left(\psi(b), \varphi(b)\right) \leq \psi\left(b\right).$$

Using Definition 3.3, we get

$$\psi\left(b\right) \leq h\left\{\psi\left(b\right)\right\} \leq \mathscr{F}\left(\psi(b), \varphi(b)\right) \leq \psi\left(b\right).$$

This gives

$$\mathscr{F}\left(\psi(b), \varphi(b)\right) = \psi(b).$$

Using by (2) of Definition 1.4, we get

$$\psi\left(b\right) = 0 \text{ or } \varphi\left(b\right) = 0.$$

Hence, $b = 0$.
i.e., $\lim\limits_{k\to\infty} \mu(\mathscr{C}_k) = 0$.

Since $\mathscr{C}_k \supseteq \mathscr{C}_{k+1}$ and $\mathscr{C}_k \supseteq \mathscr{T}\mathscr{C}_k$ forall $k \geq 1$. Then from (1.1) concluding that $\mathscr{C}_\infty = \bigcap_{k=1}^{\infty} \mathscr{C}_k$ is nonempty, convex, closed set, invariant under $\mathscr{T}$ also belong to $ker\mu$.

Thus, from Schauder's FPT (Theorem 1.1), $\mathscr{T}$ has a fixed point in $\mathscr{C}$. Hence, the completed proof.

**Theorem 2.2** *Suppose that $\mathscr{G}$ be a $\mathcal{NBCCS}$ of a Banach space $\mathscr{B}$ and, the function $\mathscr{T}:\mathscr{G}\to\mathscr{G}$ is a continuous such that*

$$m\psi\left(\mu\left(\mathscr{T}X\right)\right)\leq\mathscr{F}\left(\psi(\mu\left(X\right)),\varphi\left(\mu\left(X\right)\right)\right) \tag{2.2}$$

*where $X\subset\mathscr{G}$ and $\mu$ is an arbitrary MNC and $\psi\in\Psi$, $\varphi\in\Psi$ and $\mathscr{F}\in C$. So that $\mathscr{T}$ has atleast one fixed point in $\mathscr{G}$.*

***Proof*** Putting $h(t)=mt,\ m\geq 1$ in the Theorem 2.1. We obtain, the above Theorem.

**Theorem 2.3** *Suppose that $\mathscr{G}$ be a $\mathcal{NBCCS}$ of a Banach space $\mathscr{B}$ and, the function $\mathscr{T}:\mathscr{G}\to\mathscr{G}$ is a continuous such that*

$$\psi\left(\mu\left(\mathscr{T}X\right)\right)\leq K\psi(\mu\left(X\right)) \tag{2.3}$$

*where $X\subset\mathscr{G}$ and $\mu$ is an arbitrary MNC and $\psi\in\Psi$. So that $\mathscr{T}$ has atleast one fixed point in $\mathscr{G}$.*

***Proof*** Putting $\mathscr{F}\left(\psi(\mu\left(X\right)),\varphi\left(\mu\left(X\right)\right)\right)=nx,\ 0<n<1$ and $K=\frac{n}{m}$ in Theorem 2.2 we get, the Theorem 2.3.

**Corollary 2.1** *Suppose that $\mathscr{G}$ be a $\mathcal{NBCCS}$ of a Banach space $\mathscr{B}$ and, the function $\mathscr{T}:\mathscr{G}\to\mathscr{G}$ is a continuous such that*

$$\mu\left(\mathscr{T}X\right)\leq K\mu\left(X\right) \tag{2.4}$$

*where $X\subset\mathscr{G}$ and $\mu$ is an arbitrary MNC. So that $\mathscr{T}$ has atleast one fixed point in $\mathscr{G}$.*

***Proof*** Setting $\psi(t)=(1-k_1)t;\ 0\leq k_1\leq 1$ in Theorem 2.3 We obtain, the above Corollary.

**Theorem 2.4** *Suppose that $\mathscr{G}$ be a $\mathcal{NBCCS}$ of a Banach space $\mathscr{B}$ and, the function $\mathscr{T}:\mathscr{G}\to\mathscr{G}$ is a continuous such that*

$$\psi\left(\mu\left(\mathscr{T}X\right)\right)\leq\mathscr{F}\left(\psi(\mu\left(X\right)),\varphi\left(\mu\left(X\right)\right)\right) \tag{2.5}$$

*where $X\subset\mathscr{G}$ and $\mu$ is an arbitrary MNC and $\psi\in\Psi$, $\varphi\in\Psi$ and $\mathscr{F}\in C$. So that $\mathscr{T}$ has atleast one fixed point in $\mathscr{G}$.*

***Proof*** Assume that a sequence $\{\mathscr{C}_k\}_{k=1}^{\infty}$ with $\mathscr{C}_0=\mathscr{C}$ and $\mathscr{C}_k=Conv(\mathscr{T}\mathscr{C}_{k-1})$ for $k\geq 1$. We have, $\mathscr{T}\mathscr{C}_0=\mathscr{T}\mathscr{C}\subseteq\mathscr{C}_0=\mathscr{C}_0$, $\mathscr{C}_1=Conv(\mathscr{T}\mathscr{C}_0)\subseteq\mathscr{C}=\mathscr{C}_0$; therefore, by proceeding in the same manner gives $\mathscr{C}_0\supseteq\mathscr{C}_1\supseteq\mathscr{C}_2\supseteq\mathscr{C}_3\supseteq\ldots\supseteq\mathscr{C}_k\supseteq\mathscr{C}_{k+1}\supseteq\ldots.$

If there exists $k_0 \in \mathbb{N}$ satisfying $\mathscr{F}\left(\psi(\mu(\mathscr{C}_{k_0})), \varphi(\mu(\mathscr{C}_{k_0}))\right) = 0$. Then $\mu(\mathscr{C}_{k_0}) = 0$, so $\mathscr{C}_{k_0}$ is a relatively compact set and $\mathscr{T}\mathscr{C}_{k_0} \subseteq Conv(\mathscr{T}\mathscr{C}_{k_0}) = \mathscr{C}_{k_0+1}$. Then, Theorem 1.1 (Schauder's) yeids that $\mathscr{T}$ has a fixed point in $\mathscr{C}$.

Again, if $\mathscr{F}\,(\mu(\mathscr{C}_k), \phi(\mu(\mathscr{C}_k))) > 0,\ k \in \mathbb{N}$.

By the Eq. 2.5, we now have

$$\begin{aligned}
&\psi\,(\mu(\mathscr{C}_{k+1})) \\
&= \psi\,(\mu(Conv(\mathscr{T}\mathscr{C}_k))) \\
&= \psi\,(\mu(\mathscr{T}\mathscr{C}_k)) \\
&\leq \mathscr{F}\,(\psi(\mu(\mathscr{C}_k)), \varphi(\mu(\mathscr{C}_k))) \\
&\leq \psi(\mu(\mathscr{C}_k)).
\end{aligned}$$

Since $\psi$ is a continuous and strictly increasing mapping, we obtain

$$\mu(\mathscr{C}_{k+1}) \leq \mu(\mathscr{C}_k).$$

Then, $\{\mu(\mathscr{C}_k)\}$ is a bounded below and decreasing.
So, it is converges to $b = \inf\{\mathscr{C}_k\}$.

If possible let $b > 0$. By using (2.5), we get

$$\begin{aligned}
\psi\,(\mu(\mathscr{C}_{k+1})) &= \psi\,(\mu(Conv(\mathscr{T}\mathscr{C}_k))) \\
&\leq \mathscr{F}\,(\psi(\mu(\mathscr{C}_k)), \varphi(\mu(\mathscr{C}_k)))\,.
\end{aligned}$$

As $k \to \infty$, we get

$$\psi\,(b) \leq \mathscr{F}\,(\psi(b), \varphi(b)) \leq \psi\,(b)\,.$$

This gives

$$\mathscr{F}\,(\psi(b), \varphi(b)) = \psi(b).$$

Using by (2) of Definition 1.4, we get

$$\psi\,(b) = 0 \ or \ \varphi\,(b) = 0.$$

Hence, $b = 0$.
i.e., $\lim\limits_{k\to\infty} \mu(\mathscr{C}_k) = 0$.

Since $\mathscr{C}_k \supseteq \mathscr{C}_{k+1}$ and $\mathscr{C}_k \supseteq \mathscr{T}\mathscr{C}_k$ forall $k \geq 1$. Then from (1.1) concluding that $\mathscr{C}_\infty = \bigcap_{k=1}^{\infty} \mathscr{C}_k$ is nonempty, convex, closed set, invariant under $\mathscr{T}$ also belong to $ker\,\mu$.
Thus, from Schauder's FPT (Theorem 1.1), $\mathscr{T}$ has a fixed point in $\mathscr{C}$. Hence, the completed proof.

**Theorem 2.5** *Suppose that $\mathscr{G}$ be a $\mathcal{NBCCS}$ of a Banach space $\mathscr{B}$ and, the function $\mathscr{T}: \mathscr{G} \to \mathscr{G}$ is a continuous such that*

$$\psi\left(\mu\left(\mathscr{T}X\right)\right) \leq K\psi(\mu\left(X\right)) - \varphi\left(\mu\left(X\right)\right) \tag{2.6}$$

*where* $X \subset \mathscr{G}$ *and* $\mu$ *is an arbitrary MNC and* $\psi \in \Psi,\ \varphi \in \Psi$ *. So that* $\mathscr{T}$ *has atleast one fixed point in* $\mathscr{G}$*.*

***Proof*** Putting $\mathscr{F}\left(\psi(\mu\left(X\right)), \varphi\left(\mu\left(X\right)\right)\right) = K\psi(\mu\left(X\right)) - \varphi\left(\mu\left(X\right)\right),\ 0 < K < 1$ in the Theorem 2.4. We obtain, the above Theorem.

**Theorem 2.6** *Suppose that* $\mathscr{G}$ *be a* $\mathcal{NBCCS}$ *of a Banach space* $\mathscr{B}$ *and, the function* $\mathscr{T}: \mathscr{G} \to \mathscr{G}$ *is a continuous such that*

$$\psi\left(\mu\left(\mathscr{T}X\right)\right) \leq K\psi(\mu\left(X\right)) \tag{2.7}$$

*where* $X \subset \mathscr{G}$ *and* $\mu$ *is an arbitrary MNC and* $\psi \in \Psi$*. So that* $\mathscr{T}$ *has atleast one fixed point in* $\mathscr{G}$*.*

***Proof*** Putting $\varphi\left(\mu\left(X\right)\right) = 0$ in Theorem 2.5 we get, the Theorem 2.6.

**Corollary 2.2** *Suppose that* $\mathscr{G}$ *be a* $\mathcal{NBCCS}$ *of a Banach space* $\mathscr{B}$ *and, the function* $\mathscr{T}: \mathscr{G} \to \mathscr{G}$ *is a continuous such that*

$$\mu\left(\mathscr{T}X\right) \leq K\mu\left(X\right) \tag{2.8}$$

*where* $X \subset \mathscr{G}$ *and* $\mu$ *is an arbitrary MNC. So that* $\mathscr{T}$ *has atleast one fixed point in* $\mathscr{G}$*.*

***Proof*** Setting $\psi(t) = t$ in Theorem 2.3 We obtain, the above Corollary.

## 3 Coupled Fixed Point Theory

**Definition 3.1** ([6]) Denote by $\mathcal{Q}$ the class of all $\mathbb{Q}: \mathbb{R}_+ \to \mathbb{R}$ with the below axioms:
$(a)$ $\mathbb{Q}$ is non-decreasing.
$(b)$ $\mathbb{Q}$ is continuous.
$(c)$ $\lim_{k\to\infty} \mathbb{Q}(t_k) = 0 \Leftrightarrow \lim_{k\to\infty} t_k = 0$ for each sequence $\{t_k\} \subset (0, \infty)$.

**Definition 3.2** ([6]) Let $\Delta$ be the family of all function (known as comparison function) $\delta: \mathbb{R}_+ \to \mathbb{R}$ with the below axioms:
$(a)$ $\delta$ is monotone increasing.
$(b)$ $\lim_{q\to\infty} \delta^q(z) = 0$, for all $z > 0$.

**Definition 3.3** ([11]) An element $(\mathscr{C}, \mathscr{D}) \in \mathscr{Y} \times \mathscr{Y}$ is called a coupled fixed point of a mapping $\mathscr{T}: \mathscr{Y} \times \mathscr{Y} \to \mathscr{Y}$ if $\mathscr{T}(\mathscr{C}, \mathscr{D}) = \mathscr{C}$ and $\mathscr{T}(\mathscr{D}, \mathscr{C}) = \mathscr{D}$.

**Theorem 3.1** ([1]) *Suppose* $\mu_1, \mu_2, \ldots, \mu_n$ *is the MNC in* $\mathscr{B}_1, \mathscr{B}_2, \ldots, \mathscr{B}_n$ *respectively. Moreover, suppose the function* $\mathscr{L} : \mathbb{R}_+^n \to \mathbb{R}_+$ *is convex and* $\mathscr{J}(y_1, y_2, \ldots, y_n) = 0 \Leftrightarrow y_t = 0$ *for* $t = 1, 2, \ldots, n$ *then* $\mu(\mathscr{L}) = \mathscr{J}(\mu_1(\mathscr{L}_1), \mu_2(\mathscr{L}_2), \ldots, \mu_n(\mathscr{L}_n))$ *define a MNC in* $\mathscr{B}_1, \mathscr{B}_2, \ldots, \mathscr{B}_n$*, where* $\mathscr{L}_t$ *denotes the natural projection of* $\mathscr{L}$ *into* $\mathscr{B}_t$ *for* $t = 1, 2, \ldots, n$.

**Example 3.1** ([1]) Let $\mu$ be a MNC on $\mathscr{B}$. Define $\mathscr{J}(\mathscr{C}, \mathscr{D}) = \mathscr{C} + \mathscr{D}; \mathscr{C}, \mathscr{D} \in \mathbb{R}_+$. Then $\mathscr{J}$ has all the propertiesmentioned in Theorem 3.1. Hence, $\tilde{\mu}(\mathcal{X}) = \mu(\mathscr{L}_1) + \mu(\mathscr{L}_2)$ is a MNC in the space $\mathscr{B} \times \mathscr{B}$, where $\mathscr{L}_t$, $t = 1, 2$ denote the natural projections of $\mathscr{L}$.

**Theorem 3.2** *Let* $\mathscr{C}$ *be a NBCC subset of a Banach Space* $\mathscr{B}$ *and let* $\mathscr{T} : \mathscr{C} \times \mathscr{C} \to \mathscr{C}$ *be a continuous operator such that*

$$\begin{aligned} &\mathbb{Q}[\mathscr{F}\{\mu(\mathscr{T}(s_1 \times s_2)), \phi(\mu(\mathscr{T}(s_1 \times s_2)))\}] \\ &\quad \le \frac{1}{2}\delta[\mathbb{Q}[\mathscr{F}\{\mu(s_1), \phi(\mu(s_1))\} + \mathscr{F}\{\mu(s_2), \phi(\mu(s_2))\}]] \end{aligned} \tag{3.1}$$

*for all* $s_1, s_2 \subseteq \mathscr{C}$*, where* $\phi : \mathbb{R}_+ \to \mathbb{R}_+$ *is non-decreasing continuous function and* $\mu$ *is an arbitrary MNC. In addition, we assume* $\mathbb{Q}(\mathscr{Y} + \mathscr{Z}) \le \mathbb{Q}(\mathscr{Y}) + \mathbb{Q}(\mathscr{Z})$; $\mathscr{Y}, \mathscr{Z} \ge 0$ *and* $\phi(\mathscr{Y} + \mathscr{Z}) \le \phi(\mathscr{Y}) + \phi(\mathscr{Z})$; $\mathscr{Y}, \mathscr{Z} \ge 0$. *Then* $\mathscr{T}$ *has at least a couple* $\mathbb{FP}$ *in* $\mathscr{C}$.

***Proof*** Consider a mapping $\mathscr{T}^{cf} : \mathscr{C} \times \mathscr{C} \to \mathscr{C} \times \mathscr{C}$ by $\mathscr{T}^{cf}(\mathscr{Y}, \mathscr{Z}) = (\mathscr{T}(\mathscr{Y}, \mathscr{Z}), \mathscr{T}(\mathscr{Z}, \mathscr{Y}))$; $\mathscr{Y}, \mathscr{Z} \in \mathscr{C}$. It is trivial that $\mathscr{T}^{cf}$ is continuous.

Let $s \subseteq \mathscr{C} \times \mathscr{C}$ be nonempty. We have $\tilde{\mu}(s) = \mu(s_1) + \mu(s_2)$ is an MNC, where $s_1, s_2$ are the natural projections of $s$ into $\mathscr{B}$.

We get,

$$\begin{aligned} &\mathbb{Q}[\mathscr{F}\left\{\tilde{\mu}(\mathscr{T}^{cf}(s)), \phi(\tilde{\mu}(\mathscr{T}^{cf}(s)))\right\}] \\ &\leqslant \mathbb{Q}[\mathscr{F}\{\tilde{\mu}(\mathscr{T}(s_1 \times s_2) \times \mathscr{T}(s_2 \times s_1)), \phi(\tilde{\mu}(\mathscr{T}(s_1 \times s_2) \times \mathscr{T}(s_2 \times s_1)))\}] \\ &= \mathbb{Q}[\mathscr{F}\{\mu(\mathscr{T}(s_1 \times s_2)) + \mu(\mathscr{T}(s_2 \times s_1)), \phi(\mu(\mathscr{T}(s_1 \times s_2)) + \mu(\mathscr{T}(s_2 \times s_1)))\}] \\ &\le \mathbb{Q}[\mathscr{F}\{\mu(\mathscr{T}(s_1 \times s_2)) + \mu(\mathscr{T}(s_2 \times s_1)), \phi(\mu(\mathscr{T}(s_1 \times s_2))) + \phi(\mu(\mathscr{T}(s_2 \times s_1)))\}] \\ &\le \mathbb{Q}[\mathscr{F}\{\mu(\mathscr{T}(s_1 \times s_2)), \phi(\mu(\mathscr{T}(s_1 \times s_2)))\} + \mathscr{F}\{\mu(\mathscr{T}(s_2 \times s_1)), \phi(\mu(\mathscr{T}(s_2 \times s_1)))\}] \\ &\le \mathbb{Q}[\mathscr{F}\{\mu(\mathscr{T}(s_1 \times s_2)), \phi(\mu(\mathscr{T}(s_1 \times s_2)))\}] + \mathbb{Q}[\mathscr{F}\{\mu(\mathscr{T}(s_2 \times s_1)), \phi(\mu(\mathscr{T}(s_2 \times s_1)))\}] \\ &\le \frac{1}{2}\delta[\mathbb{Q}[\mathscr{F}\{\mu(s_1), \phi(\mu(s_1))\} + \mathscr{F}\{\mu(s_2), \phi(\mu(s_2))\}]] \\ &+ \frac{1}{2}\delta[\mathbb{Q}[\mathscr{F}\{\mu(s_2), \phi(\mu(s_2))\} + \mathscr{F}\{\mu(s_1), \phi(\mu(s_1))\}]] \\ &\le \delta[\mathbb{Q}[\mathscr{F}\{\mu(s_1), \phi(\mu(s_1))\} + \mathscr{F}\{\mu(s_2), \phi(\mu(s_2))\}]] \\ &= \delta[\mathbb{Q}[\mathscr{F}\{\mu(s_1) + \mu(s_2), \phi(\mu(s_1)) + \phi(\mu(s_2))\}]] \\ &= \delta[\mathbb{Q}[\mathscr{F}\{\tilde{\mu}(s), \phi(\tilde{\mu}(s))\}] \end{aligned}$$

By Theorem 2.1, we conclude that $\mathscr{T}^{cf}$ has minimum of one fixed point in $\mathscr{C} \times \mathscr{C}$. That is, $\mathscr{T}$ has minimum of one coupled fixed point.

# 4 Application

## 4.1 *MNC on* $\bar{\mathbb{C}}([0, \mathbb{T}])$

Suppose $\mathscr{B} = \bar{\mathbb{C}}(\bar{\mathbb{J}})$ is the space of all real valued continuous function defined on $\bar{\mathbb{J}}$, where $\bar{\mathbb{J}} = [0, \mathbb{T}]$. Then,

$$\| \mathscr{T} \| = \sup \left\{ |\mathscr{T}(x)| : x \in \bar{\mathbb{J}} \right\}, \mathscr{T} \in \mathscr{B}.$$

Let $\mathscr{F}(\neq \phi) \subseteq \mathscr{B}$ be bounded. For $\varepsilon > 0$ and $\mathscr{T} \in \mathscr{F}$ such that modulus of the continuity of $\mathscr{T}$ is defined as

$$\mu(\mathscr{T}, \varepsilon) = \sup \left\{ |\mathscr{T}(z_1) - \mathscr{T}(z_2)| : z_1, z_2 \in \bar{\mathbb{J}}, |z_1 - z_2| \leq \varepsilon \right\}.$$

Moreover, we set

$$\mu(\mathscr{T}, \varepsilon) = \sup \{ \mu(\mathscr{T}, \varepsilon) : \mathscr{T} \in \mathscr{F} \}; \ \mu_0(\mathscr{F}) = \lim_{\varepsilon \to 0} \mu(\mathscr{F}, \varepsilon).$$

Here, the mapping $\mu_0$ is known as $MNC$ in $\mathscr{B}$, with $\mathscr{Q}(\mathscr{F}) = \frac{1}{2}\mu_0(\mathscr{F})$ (see [1]) as the Hausdorff $MNC$ $\mathscr{Q}$.

## 4.2 *Existence Results of Functional Integral Equation*

In this section, we consider the following functional integral equation is given by

$$u(t) = u_0^\alpha + f(t, u(t), u(pt)) + \int_0^{qt} g_1(t, s, u(s))ds + \int_0^t g_2(t, s, u(s))ds \quad (4.1)$$

where $t \in \bar{\mathbb{J}} = [0, \mathbb{T}]$, $0 < p, q, \alpha < 1$ and $u(0) = u_0$.

Also, for $Re(\alpha) > 0$ and $[a, b] \subseteq \mathbb{R}$, we define the Riemann-Liouville fractional integral of a continuous function $f$ as follows

$$\left(I_{a+}^\alpha f\right)(x) = \frac{1}{\Gamma(\alpha)} \int_a^x \frac{f(s)}{(x-s)^{1-\alpha}} ds, x > a.$$

Putting $a = o$,

$$\left(I_{0+}^\alpha f\right)(x) = \frac{1}{\Gamma(\alpha)} \int_0^x \frac{f(s)}{(x-s)^{1-\alpha}} ds.$$

By defining

$$(G_1 u)(t) = \int_0^{qt} g_1(t, s, u(s))ds \quad (4.2)$$

$$(G_2u)(t) = \int_0^t g_2(t, s, u(s))ds \tag{4.3}$$

In this part, we study the following fractional integral equation:

$$u(t) = u_0^\alpha + f(t, u(t), u(pt)) + [I_{0+}^\alpha((G_1u)(s) + (G_2u)(s))](t) \tag{4.4}$$

Let define

$$\mathscr{Q}_{r_0} = \{u \in \mathscr{B} :\| u \|\leq r_0\}$$

and

$$\Delta = \{(t, s) : 0 \leq s \leq t\}, \Delta_q = \{(t, s) : 0 \leq s \leq qt\}$$

Assume that

(H1) $f : \bar{\mathbb{J}} \times \mathbb{R} \times \mathbb{R} \to \mathbb{R}$ is continuous function and there exists constants $a, b \geq 0$ such that

$$|f(t, x, y) - f(t, z, w)| \leq a |x - z| + b |y - w|, forall t \in \bar{\mathbb{J}} and x, y, z, w \in \mathbb{R}$$

Also,

$$\hat{f} = \sup\left\{|f(t, 0, 0)| : t \in \bar{\mathbb{J}}\right\}.$$

(H2) $g_i : \bar{\mathbb{J}} \times \bar{\mathbb{J}} \times \mathbb{R} \to \mathbb{R}$, $i = 1, 2$ are continuous function and there exists $a_1 : \Delta_q \to \mathbb{R}_+$ and $a_2 : \Delta \to \mathbb{R}_+$ such that

$$b_1(t) = \int_0^{qt} a_1(t, s)ds \in \bar{\mathbb{C}}(\bar{\mathbb{J}}), b_2(t) = \int_0^t a_2(t, s)ds \in \bar{\mathbb{C}}(\bar{\mathbb{J}})$$

and

$$|g_i(t, s, x)| \leq a_i(t, s)(1 + |x|).$$

(H3) There exists a positive solution $r_0$ such that

$$u_0^\alpha + (a + b)r_0 + \hat{f} + \frac{\mathbb{T}^\alpha}{\Gamma(\alpha + 1)}(1 + r_0) \| b_1 + b_2 \|_{\bar{\mathbb{C}}}\leq r_0.$$

**Theorem 4.1** *If conditions* $(H1) - (H3)$ *satisfied, so the Eq.(4.1) has atleast a solution in* $\mathscr{B} = \bar{\mathbb{C}}(\bar{\mathbb{J}})$.

***Proof*** We take the operator $\mathscr{D} : \mathscr{B} \to \mathscr{B}$ as follows:

$$(\mathscr{D}u)(t) = u_0^\alpha + f(t, u(t), u(pt)) + [I_{0+}^\alpha((G_1u)(s) + (G_2u)(s))](t).$$

**Step 1**: We show that $\mathscr{D}$ maps $\mathscr{Q}_{r_0}$ into $\mathscr{Q}_{r_0}$. Let $u \in \mathscr{Q}_{r_0}$. We now have

$$\begin{aligned}
&|(\mathscr{D}u)(t)| \\
&\leq |u_0^\alpha| + |f(t, u(t), u(pt))| + [I_{0+}^\alpha(|(G_1u)(s)| + |(G_2u)(s)|)](t) \\
&\leq |u_0^\alpha| + |f(t, u(t), u(pt)) - f(t, 0, 0)| + |f(t, 0, 0)| + [I_{0+}^\alpha(|(G_1u)(s)| + |(G_2u)(s)|)](t) \\
&\leq u_0^\alpha + a\,|u(t)| + b\,|u(pt)| + \hat{f} + [I_{0+}^\alpha(|(G_1u)(s)| + |(G_2u)(s)|)](t).
\end{aligned}$$

Also,

$$\begin{aligned}
&[I_{0+}^\alpha(|(G_1u)(s)| + |(G_2u)(s)|)](t) \\
&= \left[I_{0+}^\alpha\left(\left|\int_0^{qs} g_1(s, \tau, u(\tau))d\tau\right| + \left|\int_0^{s} g_2(s, \tau, u(\tau))d\tau\right|\right)\right](t) \\
&\leq \left[I_{0+}^\alpha\left(\int_0^{qs} |g_1(s, \tau, u(\tau))|\,d\tau + \int_0^{s} |g_2(s, \tau, u(\tau))|\,d\tau\right)\right](t) \\
&\leq \left[I_{0+}^\alpha\left(\int_0^{qs} a_1(s, \tau)(1 + |u(\tau)|)d\tau + \int_0^{s} a_2(s, \tau)(1 + |u(\tau)|)d\tau\right)\right](t) \\
&\leq \frac{\mathbb{T}^\alpha}{\Gamma(\alpha+1)}(1+ \parallel u \parallel) \parallel b_1 + b_2 \parallel_{\bar{\mathbb{C}}}.
\end{aligned}$$

Hence, $\parallel u \parallel < r_0$ gives

$$\parallel \mathscr{D}u \parallel \leq u_0^\alpha + (a+b)r_0 + \hat{f} + \frac{\mathbb{T}^\alpha}{\Gamma(\alpha+1)}(1 + r_0) \parallel b_1 + b_2 \parallel_{\bar{\mathbb{C}}} \leq r_0.$$

Due to the hypothesis (H3) $\mathscr{D}$ maps $\mathscr{Q}_{r_0}$ into $\mathscr{Q}_{r_0}$.

**Step 2**: Next claim that $\mathscr{D}$ is continuous on $\mathscr{Q}_{r_0}$.
Fix $\delta > 0$ and $u, u_1 \in \mathscr{Q}_{r_0}$ such that $\parallel u - u_1 \parallel < \delta$. We now have

$$\begin{aligned}
&|(\mathscr{D}u)(t) - (\mathscr{D}u_1)(t)| \\
&\leq |f(t, u(t), u(pt)) - f(t, u_1(t), u_1(pt))| \\
&+ \left|[I_{0+}^\alpha((G_1u)(s) + (G_2u)(s))](t) - [I_{0+}^\alpha((G_1u_1)(s) + (G_2u_1)(s))](t)\right| \\
&\leq a\,|u(t) - u_1(t)| + b\,|u(pt) - u_1(pt)| \\
&+ \left|[I_{0+}^\alpha((G_1u)(s) - (G_1u_1)(s) + (G_2u)(s) - (G_2u_1)(s))](t)\right| \\
&< a \parallel u - u_1 \parallel + b \parallel u - u_1 \parallel \\
&+ [I_{0+}^\alpha(|(G_1u)(s) - (G_1u_1)(s)| + |(G_2u)(s) - (G_2u_1)(s)|)](t).
\end{aligned}$$

Also,

$$[I_{0+}^\alpha(|(G_1u)(s) - (G_1u_1)(s)| + |(G_2u)(s) - (G_2u_1)(s)|)](t)$$

$$= \left[I_{0+}^{\alpha}\left(\left|\int_0^{qt}(g_1(t,s,u(s)) - g_1(t,s,u_1(s)))ds\right| + \left|\int_0^{t}(g_2(t,s,u(s)) - g_2(t,s,u_1(s)))ds\right|\right)\right](t)$$
$$\leq \left[I_{0+}^{\alpha}\left(\int_0^{qt}|g_1(t,s,u(s)) - g_1(t,s,u_1(s))|\,ds + \int_0^{t}|g_2(t,s,u(s)) - g_2(t,s,u_1(s))|\,ds\right)\right](t)$$
$$\leq \left[I_{0+}^{\alpha}\left(\mu_{r_0}(g_1,\delta)\int_0^{qt}ds + \mu_{r_0}(g_2,\delta)\int_0^{t}ds\right)\right](t)$$
$$\leq \left(\mu_{r_0}(g_1,\delta) + \mu_{r_0}(g_2,\delta)\right)\frac{\mathbb{T}^{\alpha+1}}{\Gamma(\alpha+1)}.$$

where

$$\mu_{r_0}(g_i,\delta) = \sup\left\{\begin{array}{c}|g_i(t,s,u) - g_i(t,s,u_1)| : |u - u_1| \leq \delta; t,s \in \bar{\mathbb{J}};\\ u,u_1 \in [-r_0,r_0];\ fori = 1,2.\end{array}\right\}.$$

Hence, $\| u - u_1 \| < \delta$ gives

$$|(\mathscr{D}u)(t) - (\mathscr{D}u_1)(t)| < (a+b)\delta + \left(\mu_{r_0}(g_1,\delta) + \mu_{r_0}(g_2,\delta)\right)\frac{\mathbb{T}^{\alpha+1}}{\Gamma(\alpha+1)}.$$

By the uniform continuity of $g_i, i = 1, 2$ on bounded subsets of $\bar{\mathbb{J}} \times \bar{\mathbb{J}} \times \mathbb{R}$, we obtain

$$\mu_{r_0}(g_1,\delta) \to 0\delta \to 0.$$

Thus,

$$|(\mathscr{D}u)(t) - (\mathscr{D}u_1)(t)| \to 0.$$

This clearly prove that $\mathscr{D}$ is continuous on $\mathscr{Q}_{r_0}$.

**Step 3**: An estimate of $\mathscr{D}$ with respect to $\mu_0$. Consider that $\Delta(\neq \emptyset) \subseteq \mathscr{Q}_{r_0}$. Let $\delta > 0$ be arbitrary, Also, choose $u \in \Delta$ with $t_1, t_2 \in \bar{\mathbb{J}}$ such that $|t_1 - t_2| \leq \delta$ and $t_2 \geq t_1$.

Now,

$$|(\mathscr{D}u)(t_1) - (\mathscr{D}u)(t_2)|$$
$$= \left|f(t_1,u(t_1),u(pt_1)) + [I_{0+}^{\alpha}((G_1u)(s) + (G_2u)(s))](t_1) - f(t_2,u(t_2),u(pt_2)) - [I_{0+}^{\alpha}((G_1u)(s) + (G_2u)(s))](t_2)\right|$$
$$\leq |f(t_1,u(t_1),u(pt_1)) - f(t_2,u(t_2),u(pt_2))|$$
$$+ \left|[I_{0+}^{\alpha}((G_1u)(s) + (G_2u)(s))](t_1) - [I_{0+}^{\alpha}((G_1u)(s) + (G_2u)(s))](t_2)\right|$$
$$\leq |f(t_1,u(t_1),u(pt_1)) - f(t_1,u(t_2),u(pt_2))| + |f(t_1,u(t_2),u(pt_2)) - f(t_2,u(t_2),u(pt_2))|$$
$$+ \left|[I_{0+}^{\alpha}((G_1u))(s)](t_1) - [I_{0+}^{\alpha}((G_1u)(s))](t_2)\right| + \left|[I_{0+}^{\alpha}((G_2u)(s))](t_1) - [I_{0+}^{\alpha}((G_2u)(s))](t_2)\right|$$
$$\leq a\,|u(t_1) - u(t_2)| + b\,|u(pt_2) - u(pt_2)| + \mu_f(r_0,\delta)$$
$$+ \left|[I_{0+}^{\alpha}((G_1u))(s)](t_1) - [I_{0+}^{\alpha}((G_1u)(s))](t_2)\right| + \left|[I_{0+}^{\alpha}((G_2u)(s))](t_1) - [I_{0+}^{\alpha}((G_2u)(s))](t_2)\right|$$
$$\leq (a+b)\mu(u,\delta) + \mu_f(r_0,\delta) + \left|[I_{0+}^{\alpha}((G_1u))(s)](t_1) - [I_{0+}^{\alpha}((G_1u)(s))](t_2)\right|$$
$$+ \left|[I_{0+}^{\alpha}((G_2u)(s))](t_1) - [I_{0+}^{\alpha}((G_2u)(s))](t_2)\right|.$$

where

$$\mu_f(r_0,\delta)=\sup\left\{\begin{matrix}|f(t_1,u(t_2),u(pt_2))-f(t_2,u(t_2),u(pt_2))|:|t_1-t_2|\le\delta;\,t_1,t_2\in\bar{\mathbb{J}};\\ 0<p<1;\,|u|\le r_0.\end{matrix}\right\}.$$

and

$$\mu(u,\delta)=\sup\left\{\,|u(t_1)-u(t_2)|\le\delta;\,|t_1-t_2|\le\delta;\,t_1,t_2\in\bar{\mathbb{J}}\,\right\}.$$

Also,

$$\begin{aligned}&\left|[I^{\alpha}_{0+}((G_1u))(s)](t_1)-[I^{\alpha}_{0+}((G_1u)(s))](t_2)\right|+\left|[I^{\alpha}_{0+}((G_2u)(s))](t_1)-[I^{\alpha}_{0+}((G_2u)(s))](t_2)\right|\\
&\le\frac{r_0+1}{\Gamma(\alpha)}\left|\int_0^{t_1}(b_1+b_2)(s)\left(\frac{1}{(t_1-s)^{1-\alpha}}-\frac{1}{(t_2-s)^{1-\alpha}}\right)ds+\int_{t_1}^{t_2}\frac{(b_1+b_2)(s)}{(t_2-s)^{1-\alpha}}ds\right|\\
&\le\frac{(r_0+1)\parallel b_1+b_2\parallel_{\bar{\mathbb{C}}}}{\Gamma(\alpha)}\left[\int_0^{t_1}\left(\frac{1}{(t_1-s)^{1-\alpha}}-\frac{1}{(t_2-s)^{1-\alpha}}\right)ds+\int_{t_1}^{t_2}\frac{1}{(t_2-s)^{1-\alpha}}ds\right]\\
&\le\frac{(r_0+1)\parallel b_1+b_2\parallel_{\bar{\mathbb{C}}}}{\Gamma(\alpha+1)}[2(t_2-t_1)^{\alpha}+t_2^{\alpha}-t_1^{\alpha}].\end{aligned}$$

Where, Assumption $(H2)$ implies that

$$\begin{aligned}|(G_1u)(s)|&\le\int_0^{qs}|g_1(s,\tau,u(\tau))|\,d\tau\\
&\le\int_0^{qs}a_1(s,\tau)(1+|u|)d\tau\\
&\le(1+r_0)\int_0^{qs}a_1(s,\tau)d\tau\le(1+r_0)b_1(s).\end{aligned}$$

Similarly

$$|(G_2u)(s)|\le(1+r_0)b_2(s).$$

As $\delta\to 0$, then $t_1\to t_2$, so, we obtain

$$\left|[I^{\alpha}_{0+}((G_1u))(s)](t_1)-[I^{\alpha}_{0+}((G_1u)(s))](t_2)\right|+\left|[I^{\alpha}_{0+}((G_2u)(s))](t_1)-[I^{\alpha}_{0+}((G_2u)(s))](t_2)\right|\to 0.$$

Hence,

$$|(\mathscr{D}u)(t_1)-(\mathscr{D}u)(t_2)|\le(a+b)\mu(u,\delta)+\mu_f(r_0,\delta).$$

i.e.

$$\mu(\mathscr{D}u,\delta)\le(a+b)\mu(u,\delta)+\mu_f(r_0,\delta).$$

Through the uniform continuity of $f$ on $\bar{\mathbb{J}}\times[-r_0,r_0]\times[-r_0,r_0]$ we have $\lim\limits_{\delta\to 0}\mu_f(r_0,\delta)\to 0$, as $\delta\to 0$.

Setting $\sup_{u\in\Delta}$ as well as $\delta \to 0$, we obtain

$$\mu_0(\mathscr{D}\Delta) \le (a+b)\mu_0(\Delta).$$

i.e.

$$\mu_0(\mathscr{D}\Delta) \le H\mu_0(\Delta).$$

Where

$$H = a+b$$

Hereby Corollary 2.2, $\mathscr{D}$ has a fixed point in $\Delta \subseteq \mathscr{Q}_{r_0}$.
i.e. the Eq. (4.1) has a solution in $\mathscr{B}$.

**Illustration 4.1** Take into consideration the following functional integral equation:

$$u(t) = (1)^{0.75} + \frac{u(t)+u(\frac{t}{3})}{16+t^2} + \int_0^{\frac{t}{4}} (t+s)u(s)ds + \int_0^t \frac{\ln(1+u^2(s))}{4(1+t+s)}ds \quad (4.5)$$

for $t \in [0, 1] = \bar{\mathbb{J}}$.

Here

$$u_0 = 1,$$

$$\alpha = 0.75,$$

$$f(t, u(t), u(pt)) = \frac{u(t)+u(\frac{t}{3})}{16+t^2},$$

$$p = \frac{1}{3},$$

$$g_1(t, s, u(s)) = (t+s)u(s),$$

$$q = \frac{1}{4},$$

and

$$g_2(t, s, u(s)) = \frac{\ln(1+u^2(s))}{4(1+t+s)}.$$

Also, it is obvious that $f$ is continuous fulfilling

$$|f(t, u(t), u(pt)) - f(t, u_1(t), u_1(pt))| \le \frac{|u(t)-u_1(t)|}{16} + \frac{|u(pt)-u_1(pt)|}{16}$$

Therefore, $a = b = \frac{1}{16}$.
Also,

$$|g_1(t, s, u(s))| \leq (t+s)\,|u(s)|$$

and

$$|g_2(t, s, u(s))| \leq \frac{\ln(1+|u^2(s)|)}{4(1+t+s)}.$$

So,

$$a_1(t, s) = (t+s)$$

and

$$a_2(t, s) = \frac{1}{4(1+t+s)}.$$

Therefore,

$$b_1(t) = \int_0^{\frac{t}{4}} (t+s)ds = \frac{t^2}{4} + \frac{t^2}{32}$$

and

$$b_2(t) = \int_0^t \frac{1}{4(1+t+s)} ds = \frac{1}{4}(\log(1+2t)).$$

Now,

$$b_1(t) + b_2(t) = \frac{t^2}{4} + \frac{t^2}{32} + \frac{1}{4}(\log(1+2t)).$$

If $\| u \| \leq r_0$, then

$$\hat{f} = \frac{r_0}{16},$$

$$|g_1(t, s, u(s))| \leq (t+s)r_0,$$

$$|g_2(t, s, u(s))| \leq \frac{\ln(1+r_0^2)}{4(1+t+s)}.$$

and

$$\| b_1 + b_2 \|_{\bar{\mathbb{C}}} = 0.55593$$

Puting these values in the inequality of assumption $(H3)$, we get

$$1 + \frac{3}{16}r_0 + \frac{0.555903}{\Gamma(1.75)}(1+r_0) \leq r_0$$
$$\implies 1 + \frac{0.555903}{\Gamma(1.75)} \leq \left(1 - \frac{3}{16} - \frac{0.555903}{\Gamma(1.75)}\right) r_0$$
$$\implies 7.729 \leq r_0.$$

However, assumption $(H3)$ is also fulfilled for $r_0 = [7.729, \infty)$.

Thus, we have achieved all of the assumptions from $(H1)$ to $(H3)$ in Theorem 4.1.
From Theorem 4.1, we can say that The Eq. (4.5) have solutions in $\mathscr{B} = \bar{\mathbb{C}}(\bar{\mathbb{J}})$.

**Illustration 4.2** Take into consideration the following functional integral equation:

$$u(t) = (2)^{0.9} + \frac{u(t) + u(\frac{t}{2})}{16 + t^2} + \int_0^{\frac{t}{8}} \frac{u(s)}{1 + 16\sqrt{t-s}} ds + \int_0^t \frac{(t-s)u(s)}{8} ds \quad (4.6)$$

for $t \in [0, 2] = \bar{\mathbb{J}}$.

Here

$$u_0 = 2,$$

$$\alpha = 0.9,$$

$$f(t, u(t), u(pt)) = \frac{u(t) + u(\frac{t}{2})}{16 + t^2},$$

$$p = \frac{1}{2},$$

$$g_1(t, s, u(s)) = \frac{u(s)}{1 + 16\sqrt{t-s}},$$

$$q = \frac{1}{8},$$

and

$$g_2(t, s, u(s)) = \frac{(t-s)u(s)}{8}.$$

Also, it is obvious that $f$ is continuous fulfilling

$$|f(t, u(t), u(pt)) - f(t, u_1(t), u_1(pt))| \le \frac{|u(t) - u_1(t)|}{16} + \frac{|u(pt) - u_1(pt)|}{16}$$

Therefore, $a = b = \frac{1}{16}$.
Also,

$$|g_1(t, s, u(s))| \le \frac{1}{16\sqrt{t-s}} |u(s)|$$

and

$$|g_2(t, s, u(s))| \le \frac{t-s}{8} |u(s)|.$$

So,

$$a_1(t,s) = \frac{1}{16\sqrt{t-s}}$$

and

$$a_2(t,s) = \frac{t-s}{8}.$$

Therefore,

$$b_1(t) = \int_0^{\frac{t}{8}} \frac{1}{16\sqrt{t-s}} ds = \frac{1}{32}(4-\sqrt{14})\sqrt{t}$$

and

$$b_2(t) = \int_0^{t} \frac{t-s}{8} ds = \frac{t^2}{16}.$$

Now,

$$b_1(t) + b_2(t) = \frac{1}{32}(4-\sqrt{14})\sqrt{t} + \frac{t^2}{16}.$$

If $\| u \| \leq r_0$, then

$$\hat{f} = \frac{r_0}{16},$$

$$|g_1(t,s,u(s))| \leq \frac{1}{16\sqrt{t-s}} r_0,$$

$$|g_2(t,s,u(s))| \leq \frac{t-s}{8} r_0.$$

and

$$\| b_1 + b_2 \|_{\bar{\mathbb{C}}} = 0.261417$$

Puting these values in the inequality of assumption $(H3)$, we get

$$\begin{aligned}
&(2)^{0.9} + \frac{2}{16} r_0 + \frac{1}{16} r_0 + \frac{(2)^{0.9}(0.261417)}{\Gamma(1.9)}(1+r_0) \leq r_0 \\
&\implies 1.86607 + \frac{0.487821}{\Gamma(1.9)} \leq \left(1 - \frac{3}{16} - \frac{0.487821}{\Gamma(1.9)}\right) r_0 \\
&\implies 7.77395 \leq r_0.
\end{aligned}$$

However, assumption $(H3)$ is also fulfilled for $r_0 = [7.77395, \infty)$.

Thus, we have achieved all of the assumptions from $(H1)$ to $(H3)$ in Theorem 4.1.

From Theorem 4.1, we can say that The Eq. (4.6) have solutions in $\mathscr{B} = \bar{\mathbb{C}}(\bar{\mathbb{J}})$.

# References


1. J. Banaś, K. Goebel, *Measure of Noncompactness in Banach Spaces*, Lecture Notes in Pure and Applied Mathematics, vol. 60 (Marcel Dekker, New York, 1980)
2. R.P. Agarwal, M. Meehan, D. O'regan, *Fixed Point Theory and Applications* (Cambridge University Press, 2001)
3. G. Darbo, Punti uniti in trasformazioni a codominio non compatto. Rendiconti del Seminario matematico della Università di Padova **24**, 84–92 (1955)
4. M.S. Khan, M. Swaleh, S. Sessa, Fixed point theorems by altering distances between the points. Bull. Australian Math. Soc. **30**(1), 1–9 (1984)
5. A.H. Ansari, Note on $\phi - \psi$ contractive type mappings and related fixed point, in *The 2nd Regional Conference on Mathematics and Applications* (Payame Noor University, 2014), pp. 377–380
6. I. Haque, J. Ali, M. Mursaleen, Solvability of implicit fractional order integral equation in $l_p(1 \leq p < \infty)$ space via generalized Darbo's fixed point theorem. J. Funct. Spaces (2022)
7. R. Arab, H.K. Nashine, N.H. Can, T.T. Binh, Solvability of functional-integral equations (fractional order) using measure of noncompactness. Adv. Diff. Equ. **2020**(1), 1–13 (2020)
8. M. Mursaleen, S. Mohiuddine, Applications of measures of noncompactness to the infinite system of differential equations in lp spaces. Nonlinear Anal.: Theory, Methods Appl. **75**(4), 2111–2115 (2012)
9. A. Das, B. Hazarika, R. Arab, M. Mursaleen, Applications of a fixed point theorem to the existence of solutions to the nonlinear functional integral equations in two variables. Rendiconti del Circolo Matematico di Palermo Ser. **2**, 68(1), 139–152 (2019)
10. A. Das, I. Suwan, B.C. Deuri, T. Abdeljawad, On solution of generalized proportional fractional integral via a new fixed point theorem. Adv. Diff. Equ. **2021**(1), 1–12 (2021)
11. S. Chang, Y. Cho, N. Huang, Coupled fixed point theorems with applications. J. Korean Math. Soc. **33**(3), 575–585 (1996)
12. S. Beloul, M. Mursaleen, A.H. Ansari, A generalization of Darbo's fixed point theorem with an application to fractional integral equations. J. Math. Ineq. **15**, 911–921 (2021)
13. A. Das, S.A. Mohiuddine, A. Alotaibi, B.C. Deuri, Generalization of Darbo-type theorem and application on existence of implicit fractional integral equations in tempered sequence spaces. Alexandria Eng. J. **61**(3), 2010–2015 (2022)
14. A. Das, B. Hazarika, P. Kumam, Some new generalization of Darbo's fixed point theorem and its application on integral equations. Mathematics **7**(3), 214 (2019)
15. A. Das, B. Hazarika, R. Arab, M. Mursaleen, Applications of a fixed point theorem to the existence of solutions to the nonlinear functional integral equations in two variables. Rendiconti del Circolo Matematico di Palermo Ser. **2**, 68(1), 139–152 (2019)
16. A. Das, B. Hazarika, M. Mursaleen, Application of measure of noncompactness for solvability of the infinite system of integral equations in two variables in $\ell_p$ $(1 < p < \infty)$. Revista de la Real Academia de Ciencias Exactas, Físicas y Naturales. Serie A. Matemáticas **113**(1), 31–40 (2019)
17. A. Das, B. Hazarika, V. Parvaneh, M. Mursaleen, Solvability of generalized fractional order integral equations via measures of noncompactness. Math. Sci. **15**(3), 241–251 (2021)
18. K. Kuratowski, Sur les espaces complets. Fundamenta mathematicae **1**(15), 301–309 (1930)
19. B. Hazarika, R. Arab, M. Mursaleen, Applications of measure of noncompactness and operator type contraction for existence of solution of functional integral equations. Compl. Anal. Oper. Theory **13**(8), 3837–3851 (2019)
20. B. Hazarika, R. Arab, M. Mursaleen, Application measure of noncompactness and operator type contraction for solvability of an infinite system of differential equations in lp-space. Filomat **33**(7), 2181–2189 (2019)
21. A. Das, V. Parvaneh, B.C. Deuri, Z. Bagheri, Application of a generalization of Darbo's fixed point theorem via Mizogochi-Takahashi mappings on mixed fractional integral equations involving $(k, s)$-Riemann-Liouville and Erd'elyi-Kober fractional integrals. Int. J. Nonlinear Anal. Appl. **13**(1), 859–869 (2022)

22. V. Parvaneh, N. Hussain, Z. Kadelburg, Generalized Wardowski type fixed point theorems via $\alpha$-admissible FG-contractions in b-metric spaces. Acta Mathematica Scientia **36**(5), 1445–1456 (2016)
23. M. Mursaleen, Differential equations in classical sequence spaces. Revista de la Real Academia de Ciencias Exactas, Físicas y Naturales. Serie A. Matemáticas **111**(2), 587–612 (2017)
24. H.K. Nashine, R. Arab, R.P. Agarwal, A.S. Haghighi, Darbo type fixed and coupled fixed point results and its application to integral equation. Periodica Mathematica Hungarica **77**(1), 94–107 (2018)
25. M. Rabbani, A. Das, B. Hazarika, R. Arab, Measure of noncompactness of a new space of tempered sequences and its application on fractional differential equations. Chaos Solit. Fractals **140**, 110221 (2020)
26. H.M. Srivastava, A. Das, B. Hazarika, S. Mohiuddine, Existence of solution for non-linear functional integral equations of two variables in Banach algebra. Symmetry **11**(5), 674 (2019)
27. H.K. Nashine, R. Arab, R.P. Agarwal, M. De la Sen, Positive solutions of fractional integral equations by the technique of measure of noncompactness. J. Ineq. Appl. **2017**(1), 1–16 (2017)
28. B.C. Deuri, Some new fixed point theorem via shifting distance functions. Appl. Appl. Math.: An Int. J. (AAM) **17**(3), 6 (2022)
29. S.A. Mohiuddine, A. Das, A. Alotaibi, Existence of solutions for nonlinear integral equations in tempered sequence spaces via generalized Darbo-Type theorem. J. Funct. Spaces (2022)
30. I. Haque, J. Ali, M. Mursaleen, Solvability of implicit fractional order integral equation in $\ell_p$ $(1 < p < \infty)$ space via generalized Darbo's fixed point theorem. J. Funct. Spaces (2022)
31. B.C. Deuri, A. Das, Solvability of fractional integral equations via Darbo's fixed point theorem. J. Pseudo-Diff. Oper. Appl. **13**(3), 1–12 (2022)
32. H.K. Nashine, A. Das, Extension of Darbo's fixed point theorem via shifting distance functions and its application. Nonlinear Anal.: Modell. Control **27**(2), 275–288 (2022)

# The Existence Solutions of an Integral Equation Using the Coupled Measure of Noncompactness

**Bhuban Chandra Deuri, Riken Kaye, and Rinchin Drema**

**Abstract** The aim of this article is to study the results of the fixed point theorem in measure of noncompactness and the coupled measure of noncompactness. Also, we tend to prove some results of the measure of noncompactness and the coupled measure of noncompactness for the family of contraction operators. Moreover, an application with an example is provided to illustrate the results.

**Keywords** Measure of noncompactness · Coupled measure of noncompactness · Integral equations · Darbo type fixed point theorem

**MSC Subject Classification No** 45P05 · 45G10 · 47H10

## 1 Introduction

One of the essential techniques in the nonlinear analysis is the idea of the measure of noncompactness (MNC) to handle difficulties in functional operator equations. Many authors have characterised this fundamental concept in mathematical sciences in various ways (see [4, 10–12, 16, 18, 28]). In [6], Aghajani et al. developed an application in functional integral equations and generalised Darbo's fixed-point theorem.

Kuratowski [20] was the first to suggest the MNC notion, which was essential to the development of the fixed point theory. Incorporating Kuratowski's MNC

---

B. C. Deuri (✉) · R. Kaye · R. Drema
Department of Mathematics, Rajiv Gandhi University, Doimukh, Arunachal Pradesh, India
e-mail: bhuban.deuri@rgu.ac.in

R. Kaye
e-mail: riken.kaye@rgu.ac.in

R. Drema
e-mail: rinchin.drema@rgu.ac.in

 
S. A. Mohiuddine et al. (eds.), *Positive Operators and Fixed-Point Theorems with Applications*, Industrial and Applied Mathematics,
https://doi.org/10.1007/978-981-96-2476-8_5

into his formulation, Darbo [3] expanded Schauder's fixed point theorem. Several authors then utilized these MNC in their research to investigate and determine numerous differential equations, integral equations, and integro-differential equations problems.

Banaś [1] axiomatic concept of the MNC, introduced in 1980, has sparked a lot of research interest in analyzing many problems.

Numerous researchers have generalized and utilized Darbo's theorem. This time, Darbo's theorem is generalized with an application. Numerous researchers and fractional order integral equations utilize MNC for the solution of multiple forms of integral equations(see [9, 15, 17, 19, 21, 22, 26]). Using C-class functions, Das et al. [23] developed several new generalizations.

In this article, we introduce the notion of coupled MNC to investigate fixed-point results that generalise Darbo's fixed-point theorem and many previous results in the literature. As an application, we show the existence of solutions to a functional integral equation in Banach space $BC(mathbbR+)$. Finally, an example is provided to illustrate the outcomes.

Let a real Banach space $(\mathscr{B}, \| \, . \, \|)$ and $B(c,r) = \{a \in \mathscr{B} : \| a - c \| \leq r\}$. If $\mathscr{X} (\neq \emptyset) \subseteq \mathscr{B}$. Then

- $\bar{\mathscr{X}}$ = the closure of $\mathscr{X}$,
- $Conv\mathscr{X}$ = the convex closure of $\mathscr{X}$,
- $\mathfrak{M}_{\mathscr{B}}$ = collection of all non-empty and bounded subsets of $\mathscr{B}$,
- $\mathfrak{N}_{\mathscr{B}}$ = collection of all relatively compact sets,
- $\mathbb{R} = (-\infty, \infty)$,
  also
- $\mathbb{R}_+ = [0, \infty)$.

The definition of a MNC is as follows: [1].

**Definition 1.1** A function $\mu : \mathfrak{M}_{\mathscr{B}} \to [0, \infty)$ is said to be a MNC in $\mathscr{B}$ if it fulfills axioms:

(i) for all $\mathscr{X} \in \mathfrak{M}_{\mathscr{B}}$, we have $\mu(\mathscr{X}) = 0$ gives $\mathscr{X}$ is relatively compact.
(ii) $\ker \mu = \{\mathscr{X} \in \mathfrak{M}_{\mathscr{B}} : \mu(\mathscr{X}) = 0\} \neq \emptyset$ and $\ker \mu \subset \mathfrak{N}_{\mathscr{B}}$.
(iii) $\mathscr{X} \subseteq \mathscr{X}_1 \implies \mu(\mathscr{X}) \leq \mu(\mathscr{X}_1)$.
(iv) $\mu\left(\bar{\mathscr{X}}\right) = \mu(\mathscr{X})$.
(v) $\mu(Conv\mathscr{X}) = \mu(\mathscr{X})$.
(vi) $\mu(\varpi\mathscr{X} + (1-\varpi)\mathscr{X}_1) \leq \varpi\mu(\mathscr{X}) + (1-\varpi)\mu(\mathscr{X}_1)$ for $\varpi \in [0, 1]$.
(vii) if $\mathscr{X}_k \in \mathfrak{M}_{\mathscr{B}}$, $\mathscr{X}_k = \bar{\mathscr{X}_k}$, $\mathscr{X}_{k+1} \subset \mathscr{X}_k$ for $k = 1, 2, 3, 4, 5, \ldots$ and $\lim_{k\to\infty} \mu(\mathscr{X}_k) = 0$ then $\bigcap_{k=1}^{\infty} \mathscr{X}_k \neq \emptyset$.

The *kernel of measure* $\mu$ is defined to be the family $ker\mu$. Since $\mu(\mathscr{X}_\infty) \leq \mu(\mathscr{X}_k)$, $\mu(\mathscr{X}_\infty) = 0$. So, $\mathscr{X}_\infty = \bigcap_{k=1}^{\infty} \mathscr{X}_k \in ker\mu$.

### *Some Important Theorem and Definition*

The following are some fundamental theorems to recall:

**Theorem 1.1** *Let $\mathscr{G}$ be a nonempty, bounded, closed, and convex subset(NBCCS) of a Banach space $\mathscr{B}$ and $\mathscr{T} : \mathscr{G} \to \mathscr{G}$ be a compact and continuous operator. Then, $\mathscr{T}$ has at least one fixed-point.*

**Theorem 1.2** (Shauder [2]) *Let $\mathscr{G}$ be a NBCCS of a Banach Space $\mathscr{B}$. If $\mathscr{T} : \mathscr{G} \to \mathscr{G}$ is a continuous and compact mapping, then it must have at least one fixed point.*

**Theorem 1.3** (Darbo [3]) *Let $\mathscr{G}$ be a NBCCS of a Banach Space $\mathscr{B}$. Assume $\mathscr{T} : \mathscr{G} \to \mathscr{G}$ is a continuous mapping with a constant $\varpi \in [0, 1)$ such that*

$$\mu(\mathscr{T}\mathscr{X}) \leq \varpi \mu(\mathscr{X}), \ \mathscr{X} \subseteq \mathscr{G}.$$

*Then $\mathscr{T}$ has a fixed point.*

The following related concepts are needed to establish an extension of Darbo's fixed point theorem:

**Definition 1.2** ([5]) A function $\psi : \mathbb{R}_+ \to \mathbb{R}_+$ is an altering distance function if the subsequent properties hold true:
$(a)$ $\psi(t) = 0 \iff t = 0$,
$(b)$ $\psi$ is continuous and strictly increasing.
where, $\psi \in \Psi$ is the set of altering distance functions.

For example,
(1) $\psi(t) = (1 - k)t, 0 \leq k < 1$.

**Definition 1.3** ([7]) A continuous function $\mathscr{F} : \mathbb{R}_+ \times \mathbb{R}_+ \to \mathbb{R}_+$ is a function of C-class, if subsequent axioms hold true:
$a$ $\mathscr{F}(p, w) \leq p$;
$b$ $\mathscr{F}(p, w) = p$ implies that either $p = 0$ or $w = 0$, $p, w \in \mathbb{R}_+$.
Noticeably, $\mathscr{F}(0, 0) = 0$. A collection of C-class of functions is symbolized as C.

For example,
(1) $\mathscr{F}(p, w) = p - \frac{w}{k+w}, \ 0 \leq k < 1$.
(2) $\mathscr{F}(p, w) = \frac{p-w}{1+w}$.

## 2 Darbo Fixed Point Theory Results

**Theorem 2.1** *Suppose that $\mathscr{G}$ be a NBCCS of a Banach space $\mathscr{B}$ and, the function $\mathscr{T} : \mathscr{G} \to \mathscr{G}$ is a continuous such that*

$$\psi\left(\mu\left(\mathscr{T}X\right)\right) \leq \mathscr{F}\left(\psi(\mu\left(X\right)), \mu\left(X\right)\right) \tag{2.1}$$

*where $X \subset \mathscr{G}$ and $\mu$ is an arbitrary MNC and $\psi \in \Psi$ and $\mathscr{F} \in C$. So that $\mathscr{T}$ has atleast one fixed point in $\mathscr{G}$.*

***Proof*** Assume that a sequence $\{\mathscr{C}_k\}_{k=1}^{\infty}$ with $\mathscr{C}_0 = \mathscr{C}$ and $\mathscr{C}_k = Conv(\mathscr{T}\mathscr{C}_{k-1})$ for $k \geq 1$. We have, $\mathscr{T}\mathscr{C}_0 = \mathscr{T}\mathscr{C} \subseteq \mathscr{C}_0 = \mathscr{C}_0$, $\mathscr{C}_1 = Conv(\mathscr{T}\mathscr{C}_0) \subseteq \mathscr{C} = \mathscr{C}_0$; therefore, by proceeding in the same manner gives $\mathscr{C}_0 \supseteq \mathscr{C}_1 \supseteq \mathscr{C}_2 \supseteq \mathscr{C}_3 \supseteq \ldots \supseteq \mathscr{C}_k \supseteq \mathscr{C}_{k+1} \supseteq \ldots$.

If there exists $k_0 \in \mathbb{N}$ satisfying $\mathscr{F}\left(\psi(\mu(\mathscr{C}_{k_0})), \mu(\mathscr{C}_{k_0})\right) = 0$. Then $\mu(\mathscr{C}_{k_0}) = 0$, so $\mathscr{C}_{k_0}$ is a relatively compact set and $\mathscr{T}\mathscr{C}_{k_0} \subseteq Conv(\mathscr{T}\mathscr{C}_{k_0}) = \mathscr{C}_{k_0+1}$. Then, Theorem 1.2 (Schauder's) yeids that $\mathscr{T}$ has a fixed point in $\mathscr{C}$.

Again, if $\mathscr{F}\left(\psi(\mu(\mathscr{C}_k)), \mu(\mathscr{C}_k)\right) > 0,\ k \in \mathbb{N}$.

By the Eq. 2.1, we now have

$$\begin{aligned}
&\psi\left(\mu(\mathscr{C}_{k+1})\right) \\
&= \psi\left(\mu(Conv(\mathscr{T}\mathscr{C}_k))\right) \\
&= \psi\left(\mu(\mathscr{T}\mathscr{C}_k)\right) \\
&\leq \mathscr{F}\left(\psi(\mu(\mathscr{C}_k)), \mu(\mathscr{C}_k)\right) \\
&\leq \psi(\mu(\mathscr{C}_k)).
\end{aligned}$$

Since $\psi$ is a continuous and strictly increasing mapping, we obtain

$$\mu(\mathscr{C}_{k+1}) \leq \mu(\mathscr{C}_k).$$

Then, $\{\mu(\mathscr{C}_k)\}$ is a bounded below and decreasing.
So, it is converges to $b = \inf\{\mathscr{C}_k\}$.

If possible let $b > 0$. By using (2.1), we get

$$\begin{aligned}
\psi\left(\mu(\mathscr{C}_{k+1})\right) &= \psi\left(\mu(Conv(\mathscr{T}\mathscr{C}_k))\right) \\
&\leq \mathscr{F}\left(\psi(\mu(\mathscr{C}_k)), \mu(\mathscr{C}_k)\right).
\end{aligned}$$

As $k \to \infty$, we get

$$\psi\left(b\right) \leq \mathscr{F}\left(\psi(b), b\right) \leq \psi\left(b\right).$$

This gives

$$\mathscr{F}\left(\psi(b), b\right) = \psi(b).$$

Using by (2) of Definition 1.3, we get

$$\psi\left(b\right) = 0 \text{ or } b = 0.$$

Hence, $b = 0$.
i.e., $\lim\limits_{k\to\infty} \mu(\mathscr{C}_k) = 0$.

Since $\mathscr{C}_k \supseteq \mathscr{C}_{k+1}$ and $\mathscr{C}_k \supseteq \mathscr{T}\mathscr{C}_k$ forall $k \geq 1$. Then from (1.1) concluding that $\mathscr{C}_\infty = \bigcap_{k=1}^{\infty} \mathscr{C}_k$ is nonempty, convex, closed set, invariant under $\mathscr{T}$ also belong to $ker\,\mu$.
Thus, from Schauder's FPT (Theorem 1.2), $\mathscr{T}$ has a fixed point in $\mathscr{C}$. Hence, the completed proof. □

**Theorem 2.2** *Suppose that $\mathscr{G}$ be a NBCCS of a Banach space $\mathscr{B}$ and, the function $\mathscr{T}:\mathscr{G}\to\mathscr{G}$ is a continuous such that*

$$\psi\left(\mu\left(\mathscr{T}X\right)\right) \leq K\psi(\mu\left(X\right)) \tag{2.2}$$

*where $X \subset \mathscr{G}$ and $\mu$ is an arbitrary MNC and $\psi \in \Psi$. So that $\mathscr{T}$ has atleast one fixed point in $\mathscr{G}$.*

***Proof*** Putting $\mathscr{F}\left(\psi(\mu\left(X\right)), \mu\left(X\right)\right) = K\psi(\mu\left(X\right)),\ 0 < K < 1$ in the Theorem 2.1. We obtain, the above Theorem. □

**Corollary 2.1** *Suppose that $\mathscr{G}$ be a NBCCS of a Banach space $\mathscr{B}$ and, the function $\mathscr{T}:\mathscr{G}\to\mathscr{G}$ is a continuous such that*

$$\mu\left(\mathscr{T}X\right) \leq K\mu\left(X\right) \tag{2.3}$$

*where $X \subset \mathscr{G}$ and $\mu$ is an arbitrary MNC. So that $\mathscr{T}$ has atleast one fixed point in $\mathscr{G}$.*

***Proof*** Setting $\psi(t) = t$ in Theorem 2.2. We obtain, the above Corollary. □

## 3 Coupled Measure of Noncompactness

The definition of a coupled MNC is as follows: [1].

**Definition 3.1** Let $\mathscr{B}$ be a Banach space and a function $\mu : \mathfrak{M}^2_{\mathscr{B}} \to [0, \infty)$ is said to be a coupled MNC in $\mathscr{B}$ if it fulfills axioms:

(i) $\ker\mu = \left\{(\mathscr{X}^1, \mathscr{X}^2) \in \mathfrak{M}^2_{\mathscr{B}} : \mu\left(\mathscr{X}^1, \mathscr{X}^2\right) = 0\right\} \neq \emptyset$ and $\ker\mu \subset \mathfrak{N}_{\mathscr{B}}$.
(ii) For every $(\mathscr{X}^1, \mathscr{X}^2) \in \mathfrak{M}^2_{\mathscr{B}}$, we have $\mu(\mathscr{X}^1, \mathscr{X}^2) = 0$ gives $(\mathscr{X}^1, \mathscr{X}^2)$ is relatively compact.
(iii) For each $((\mathscr{X}^1, \mathscr{X}^2), (\mathscr{X}^1_1, \mathscr{X}^2_1)) \in \mathfrak{M}^2_{\mathscr{B}} \times \mathfrak{M}^2_{\mathscr{B}}$ and $(\mathscr{X}^1, \mathscr{X}^2) \subseteq (\mathscr{X}^1_1, \mathscr{X}^2_1)$, where $\mathscr{X}^1 \subseteq \mathscr{X}^1_1$ and $\mathscr{X}^2 \subseteq \mathscr{X}^2_1$,
we have
$(\mathscr{X}^1, \mathscr{X}^2) \subseteq (\mathscr{X}^1_1, \mathscr{X}^2_1) \implies \mu\left(\mathscr{X}^1, \mathscr{X}^2\right) \leq \mu\left(\mathscr{X}^1_1, \mathscr{X}^2_1\right)$.
(iv) For every $(\mathscr{X}^1, \mathscr{X}^2) \in \mathfrak{M}^2_{\mathscr{B}}$, $\mu\left(\bar{\mathscr{X}^1}, \bar{\mathscr{X}^2}\right) = \mu\left(\mathscr{X}^1, \mathscr{X}^2\right) = \mu\,(Conv\left(\mathscr{X}^1, \mathscr{X}^2\right))$.
(v) $\mu\left(\varpi(\mathscr{X}^1, \mathscr{X}^2) + (1-\varpi)(\mathscr{X}^1_1, \mathscr{X}^2_1)\right) \leq \varpi\mu\left(\mathscr{X}^1, \mathscr{X}^2\right) + (1-\varpi)\,\mu\left(\mathscr{X}^1_1, \mathscr{X}^2_1\right)$ for $\varpi \in [0, 1]$.

(vi) if $\{\mathscr{X}_k^1\}_0^{+\infty}$, $\{\mathscr{X}_k^2\}_0^{+\infty}$ in $\mathfrak{M}_{\mathscr{B}}$ are decreasing sequences of closed sets and $\lim_{k\to\infty} \mu\left\{\left(\mathscr{X}_k^1, \mathscr{X}_k^2\right)\right\}_0^{+\infty} = 0$, then $\left(\mathscr{X}_{+\infty}^1, \mathscr{X}_{+\infty}^2\right) = \bigcap_{k=0}^{+\infty}\left(\mathscr{X}_k^1, \mathscr{X}_k^2\right) \neq \emptyset$.

**Theorem 3.1** *Suppose that $\mathscr{G}$ be a NBCCS of a Banach space $\mathscr{B}$ and, the function $\mathscr{T}:\mathscr{G}\to\mathscr{G}$ is a continuous such that*

$$\psi\left(\mu\left(\mathscr{T}X^1, \mathscr{T}X^2\right)\right) \leq \mathscr{F}\left(\psi(\mu\left(X^1, X^2\right)), \mu\left(X^1, X^2\right)\right) \tag{3.1}$$

*where $X^1 \subset \mathscr{G}$, $X^1 \subset \mathscr{G}$ and $\mu$ is an arbitrary coupled MNC and $\psi \in \Psi$ and $\mathscr{F} \in C$. So that $\mathscr{T}$ has atleast one fixed point in $\mathscr{G}$.*

***Proof*** Taking $\mathscr{C}_0^1, \mathscr{C}_0^2 = \mathscr{G}$, $\mathscr{C}_{k+1}^1 = Conv(\mathscr{T}\mathscr{C}_k^1)$, $\mathscr{C}_{k+1}^2 = Conv(\mathscr{T}\mathscr{C}_k^2)$, for $n = 0, 1, 2, \ldots\ldots.$ We obtain $\mathscr{C}_{k+1}^1 \subseteq \mathscr{C}_k^1$, $\mathscr{C}_{k+1}^2 \subseteq \mathscr{C}_k^2$ for $n = 0, 1, \ldots.$ Therefore, $\{\mathscr{C}_k^1\}_0^{+\infty}$, $\{\mathscr{C}_k^2\}_0^{+\infty}$ are decreasing sequences of closed and convex sets.

By the Eq. 3.1, we now have

$$\begin{aligned}
&\psi\left(\mu(\mathscr{C}_{k+1}^1, \mathscr{C}_{k+1}^2)\right)\\
&= \psi\left(\mu(Conv(\mathscr{T}\mathscr{C}_k^1), Conv(\mathscr{T}\mathscr{C}_k^2))\right)\\
&= \psi\left(\mu(\mathscr{T}\mathscr{C}_k^1, \mathscr{T}\mathscr{C}_k^2)\right)\\
&\leq \mathscr{F}\left(\psi(\mu(\mathscr{C}_k^1, \mathscr{C}_k^2)), \mu(\mathscr{C}_k^1, \mathscr{C}_k^2)\right)\\
&\leq \psi(\mu(\mathscr{C}_k^1, \mathscr{C}_k^2)).
\end{aligned}$$

Since $\psi$ is a continuous and strictly increasing mapping, we obtain

$$\mu(\mathscr{C}_{k+1}^1, \mathscr{C}_{k+1}^2) \leq \mu(\mathscr{C}_k^1, \mathscr{C}_k^2).$$

Then, $\left\{\mu(\mathscr{C}_k^1, \mathscr{C}_k^2)\right\}$ is a bounded below and decreasing.
So, it is converges to $b = \inf\left\{\mathscr{C}_k^1, \mathscr{C}_k^2\right\}$.

If possible let $b > 0$. By using (3.1), we get

$$\begin{aligned}
\psi\left(\mu(\mathscr{C}_{k+1}^1, \mathscr{C}_{k+1}^2)\right) &= \psi\left(\mu(Conv(\mathscr{T}\mathscr{C}_k^1), Conv(\mathscr{T}\mathscr{C}_k^2))\right)\\
&\leq \mathscr{F}\left(\psi(\mu(\mathscr{C}_k^1, \mathscr{C}_k^2)), \mu(\mathscr{C}_k^1, \mathscr{C}_k^2)\right).
\end{aligned}$$

As $k \to \infty$, we get

$$\psi(b) \leq \mathscr{F}(\psi(b), b) \leq \psi(b).$$

This yields

$$\mathscr{F}(\psi(b), b) = \psi(b).$$

Using by (2) of Definition 1.3, we get

$$\psi(b) = 0 \ or\ b = 0.$$

Consequently, $b = 0$.
So, $\lim_{k\to\infty} \mu(\mathscr{C}_k^1, \mathscr{C}_k^2) = 0$.

Since $(\mathscr{C}_k^1, \mathscr{C}_k^2) \supseteq (\mathscr{C}_{k+1}^1, \mathscr{C}_{k+1}^2)$ and $(\mathscr{C}_k^1, \mathscr{C}_k^2) \supseteq (\mathscr{T}\mathscr{C}_k^1, \mathscr{T}\mathscr{C}_k^2)$ forall $k \geq 1$. Then from (1.1) concluding that $(\mathscr{C}_{+\infty}^1, \mathscr{C}_{+\infty}^2) = \bigcap_{k=0}^{+\infty}(\mathscr{C}_k^1, \mathscr{C}_k^2)$ is nonempty, convex, closed set, invariant under $\mathscr{T}$ also belong to $ker\mu$.
Thus, from Schauder's FPT (Theorem 1.2), $\mathscr{T}$ has a fixed point in $\mathscr{C}$. Hence, the completed proof. □

## 4 Application

### *4.1 Coupled MNC on $BC(\mathbb{R}_+)$*

We offer the applications of Theorem 3.1 to prove the existence of solutions of a functional integral equation in the Banach space $BC(\mathbb{R}_+)$ consisting of all real functions that are bounded and continuous on $\mathbb{R}_+$. This space is endowed with the supremum norm

$$\| x^1 \| = \sup\left\{\left|x^1(r)\right| : r \in \mathbb{R}_+\right\}.$$

We choose nonempty bounded subsets $\mathscr{X}^1$, $\mathscr{X}^2$ of $BC(\mathbb{R}_+)$ and a number $\mathbb{T} > 0$. For $\varepsilon > 0$, $x^1 \in \mathscr{X}^1$, and $x^2 \in \mathscr{X}^2$, we show the modulus of continuity of function $(x^1, x^2)$ on the interval $[0, \mathbb{T}]$ by $\mu^{\mathbb{T}}((x^1, x^2), \varepsilon)$ :

$$\mu^{\mathbb{T}}((x^1, x^2), \varepsilon) = \sup\left\{\left|x^1x^2(r) - x^1x^2(s)\right| : r, s \in [0, \mathbb{T}], |r - s| \leq \varepsilon\right\}.$$

Also,

$$\mu^{\mathbb{T}}((\mathscr{X}^1, \mathscr{X}^2), \varepsilon) = \sup\left\{\mu^{\mathbb{T}}((x^1, x^2), \varepsilon) : x^1 \in \mathscr{X}^1, x^2 \in \mathscr{X}^2\right\},$$

$$\mu_0^{\mathbb{T}}(\mathscr{X}^1, \mathscr{X}^2) = \lim_{\varepsilon\to 0} \mu^{\mathbb{T}}((x^1, x^2), \varepsilon),$$

$$\mu_0(\mathscr{X}^1, \mathscr{X}^2) = \lim_{\varepsilon\to 0} \mu^{\mathbb{T}}(x^1, x^2).$$

Moreover, for $r \in \mathbb{R}_+$, we define

$$(\mathscr{X}^1, \mathscr{X}^2)(r) = \left\{(x^1, x^2)(r) : x^1 \in \mathscr{X}^1, x^2 \in \mathscr{X}^2\right\},$$

$$\mu(\mathscr{X}^1, \mathscr{X}^2) = \mu_0(\mathscr{X}^1, \mathscr{X}^2) + \lim_{r\to+\infty} \sup diam(\mathscr{X}^1, \mathscr{X}^2)(r).$$

$$diam(\mathscr{X}^1, \mathscr{X}^2)(r) = \sup\left\{\left|x^1x^2(r) - x_1^1x_1^2(r)\right| : x^1, x_1^1 \in \mathscr{X}^1,\ x^2, x_1^2 \in \mathscr{X}^2\right\}.$$

## 4.2 The Existence Solutions of Integral Equation

In this section, we consider the following integral equation is given by

$$(x^1, x^2)(r) = p(r) + f(r, x^1(r), x^2(r)) \int_0^r v(r, s, x^1(s), x^2(s))ds, \quad r \geq 0. \quad (4.1)$$

We consider the following hypotheses:

(H1)

$$p \in BC(\mathbb{R}_+).$$

(H2) $f : \mathbb{R}_+ \times \mathbb{R}^2 \to \mathbb{R}$ is a function continuous, and there exists a continuous mapping $m : \mathbb{R}_+ \to \mathbb{R}_+$ such that

$$\left|f(r, x^1, x^2) - f(r, x_1^1, x_1^2)\right| \leq m(r) \left|x^1 x^2 - x_1^1 x_1^2\right|$$

for any $r \in \mathbb{R}_+$ and for all $x^1, x^2, x_1^1, x_1^2 \in \mathbb{R}$.

(H3) $v(r, s, x^1, x^2) = v : \mathbb{R}_+^2 \times \mathbb{R}^2 \to \mathbb{R}$ is a continuous function. Moreover, there exists $h : \mathbb{R}_+^2 \to \mathbb{R}_+$, being a continuous function on $\mathbb{R}_+^2$ and $H : \mathbb{R}_+ \to \mathbb{R}_+$, which is a continuous function and non-decreasing on $\mathbb{R}_+$ with $H(0) = 0$ and such that

$$\left|v(r, s, x^1, x^2) - v(r, s, x_1^1, x_1^2)\right| \leq h(r, s)H(\left|x^1 x^2 - x_1^1 x_1^2\right|)$$

for any $r, s \in \mathbb{R}_+$ such that $r \geq s$ and for all $x^1, x^2, x_1^1, x_1^2 \in \mathbb{R}$.

(H4) $a, b, c, d : \mathbb{R}_+ \to \mathbb{R}_+$ are the functions, defined by the formulas

$$a(r) = m(r) \int_0^r h(r, s)ds,$$

$$b(r) = |f(r, 0, 0)| \int_0^r h(r, s)ds,$$

$$c(r) = m(r) \int_0^r |v(r, s, 0, 0)|\, ds,$$

$$d(r) = |f(r, 0, 0)| \int_0^r |v(r, s, 0, 0)|\, ds,$$

are bounded on $\mathbb{R}_+$, and the functions $a(r)$ and $b(r)$ vanish at infinity, i.e., $\lim_{r \to \infty} a(r) = \lim_{r \to \infty} b(r) = 0$.

Taking into account the above hypotheses, we may define the following finite constants:

$$A = \sup \{a(r) : r \in \mathbb{R}_+\},$$

$$B = \sup \{b(r) : r \in \mathbb{R}_+\},$$

$$C = \sup \{c(r) : r \in \mathbb{R}_+\},$$

$$D = \sup \{d(r) : r \in \mathbb{R}_+\}.$$

(H5) There exists a positive solution $r_0$ such that

$$\| p \| + Ar_0 H(r_0) + BH(r_0) + Cr_0 + D \leq r_0$$

with $k < 1$, where $k = AH(r_0) + C$.

**Theorem 4.1** *According to hypotheses $(H1) - (H5)$ satisfied, relation Eq. (4.1) has at least one solution in $BC(\mathbb{R}_+)$.*

***Proof*** We define operator $V$ on $BC(\mathbb{R}_+)$ by

$$(Vx^1, Vx^2)(r) = p(r) + f(r, x^1(r), x^2(r)) \int_0^r v(r, s, x^1(s), x^2(s))ds, \quad for\ r \in \mathbb{R}_+.$$

For arbitrary $x^1, x^2 \in BC(\mathbb{R}_+)$. Observe that in view of assumptions hypotheses $(H1) - (H3)$, the functions $Vx^1$, $Vx^2$ are continuous on $\mathbb{R}_+$. Moreover, for a fixed $r \in \mathbb{R}_+$, We obtain

$$\begin{aligned}
&\left|(Vx^1, Vx^2)(r)\right| \\
&\leq |p(r)| + \left[\left|f(r, x^1(r), x^2(r)) - f(r, 0, 0)\right| + |f(r, 0, 0)|\right] \int_0^r \left|v(r, s, x^1(s), x^2(s))\right| ds \\
&\leq |p(r)| + \left[m(r)\left|x^1(r)x^2(r)\right| + |f(r, 0, 0)|\right] \int_0^r \left[\left|v(r, s, x^1(s), x^2(s)) - v(r, s, 0, 0)\right| + |v(r, s, 0, 0)|\right] ds \\
&\leq |p(r)| + \left[m(r)\left|x^1(r)x^2(r)\right| + |f(r, 0, 0)|\right] \int_0^r \left[h(r, s)H(\left|x^1(s)x^2(s)\right|) + |v(r, s, 0, 0)|\right] ds \\
&\leq \| p \| + m(r) \| x^1x^2 \| H(\| x^1x^2 \|) \int_0^r h(r, s)ds + H(\| x^1x^2 \|) |f(r, 0, 0)| \int_0^r h(r, s)ds \\
&+ m(r) \| x^1x^2 \| \int_0^r |v(r, s, 0, 0)| ds + |f(r, 0, 0)| \int_0^r |v(r, s, 0, 0)| ds \\
&= \| p \| + a(r) \| x^1x^2 \| H(\| x^1x^2 \|) + H(\| x^1x^2 \|)b(r) + c(r) \| x^1x^2 \| + d(r) \\
&\leq \| p \| + A \| x^1x^2 \| H(\| x^1x^2 \|) + H(\| x^1x^2 \|)B + C \| x^1x^2 \| + D.
\end{aligned}$$

In view of the above estimate, we infer that the function $V$ is bounded on the interval $\mathbb{R}_+$. This implies that $V \in BC(\mathbb{R}_+)$, which means that the operator $V$ transforms the space $BC(\mathbb{R}_+)$ into itself.

Apart from this, keeping in mind hypothesis $(H5)$, we deduce that there exists a number $r_0 > 0$ such that the operator $V$ transforms the ball $B_{r_0}$ into itself and $k = AH(r_0) + C < 1$.

Now, fix arbitrarily $x^1, x^2, x_1^1, x_1^2 \in B_{r_0}$. Then, for a fixed $r \in \mathbb{R}_+$, we get

$$\left|(Vx^1, Vx^2)(r) - (Vx_1^1, Vx_2^2)(r)\right|$$

$$\leq \left|f(r,x^1(r),x^2(r))\int_0^r v(r,s,x^1(s),x^2(s))ds - f(r,x_1^1(r),x_1^2(r))\int_0^r v(r,s,x^1(s),x^2(s))ds\right|$$
$$+\left|f(r,x_1^1(r),x_1^2(r))\int_0^r v(r,s,x^1(s),x^2(s))ds - f(r,x_1^1(r),x_1^2(r))\int_0^r v(r,s,x_1^1(s),x_1^2(s))ds\right|$$
$$\leq |f(r,x^1(r),x^2(r)) - f(r,x_1^1(r),x_1^2(r))|\int_0^r |v(r,s,x^1(s),x^2(s))|\,ds$$
$$+|f(r,x_1^1(r),x_1^2(r))|\left|\int_0^r v(r,s,x^1(s),x^2(s))ds - \int_0^r v(r,s,x_1^1(s),x_1^2(s))ds\right|$$
$$\leq m(r)\left|x^1(r)x^2(r) - x_1^1(r)x_1^2(r)\right|\int_0^r \left[\,|v(r,s,x^1(s),x^2(s)) - v(r,s,0,0)| + |v(r,s,0,0)|\,\right]ds$$
$$+\left[\,|f(r,x_1^1(r),x_1^2(r)) - f(r,0,0)| + |f(r,0,0)|\,\right]\int_0^r |v(r,s,x^1(s),x^2(s)) - v(r,s,x_1^1(s),x_1^2(s))|\,ds$$
$$\leq m(r)\left|x^1(r)x^2(r) - x_1^1(r)x_1^2(r)\right|\int_0^r \left[h(r,s)H(|x^1(s)x^2(s)|) + |v(r,s,0,0)|\,\right]ds$$
$$+\left[m(r)\left|x_1^1(r)x_1^2(r)\right| + |f(r,0,0)|\,\right]\int_0^r h(r,s)H(|x^1(s)x^2(s) - x_1^1(s)x_1^2(s)|)ds$$
$$\leq m(r)\left[\,|x^1(r)x^2(r)| + |x_1^1(r)x_1^2(r)|\,\right]\int_0^r h(r,s)H(|x^1(s)x^2(s)|)ds$$
$$+m(r)\left|x^1(r)x^2(r) - x_1^1(r)x_1^2(r)\right|\int_0^r |v(r,s,0,0)|\,ds$$
$$+m(r)\left|x_1^1(r)x_1^2(r)\right|\int_0^r h(r,s)H(|x^1(s)x^2(s)| + |x_1^1(s)x_1^2(s)|)ds$$
$$+|f(r,0,0)|\int_0^r h(r,s)H(|x^1(s)x^2(s)| + |x_1^1(s)x_1^2(s)|)ds$$
$$\leq 2r_0 m(r)H(r_0)\int_0^r h(r,s)ds + m(r)\left|x^1(r)x^2(r) - x_1^1(r)x_1^2(r)\right|\int_0^r |v(r,s,0,0)|\,ds$$
$$+m(r)r_0H(2r_0)\int_0^r h(r,s)ds + H(2r_0)\,|f(r,0,0)|\int_0^r h(r,s)ds$$
$$= 2r_0H(r_0)a(r) + c(r)\left|x^1(r)x^2(r) - x_1^1(r)x_1^2(r)\right| + a(r)r_0H(2r_0) + H(2r_0)b(r).$$

Hence, we get

$$diam(V\mathscr{X}^1, V\mathscr{X}^2)(r) \leq 2r_0H(r_0)a(r) + c(r)diam(\mathscr{X}^1, \mathscr{X}^2)(r) + a(r)r_0H(2r_0) + H(2r_0)b(r).$$

Joining the above estimate and hypothesis $(H4)$, we derive the following inequality

$$\limsup_{r\to+\infty} diam(V\mathscr{X}^1, V\mathscr{X}^2)(r) \leq k \limsup_{r\to+\infty} diam(\mathscr{X}^1, \mathscr{X}^2)(r).$$

where $k$ is the constant defined in hypothesis $(H5)$.

Next, let us fix arbitrarily numbers $\mathbb{T} > 0$ and $\epsilon > 0$. Choose a function $x^1 \in \mathscr{X}^1, x^2 \in \mathscr{X}^2$ and take $r_1, r_2 \in [0, \mathbb{T}]$ such that $|r_2 - r_1| \leq \epsilon$. Without the loss of generality, we may assume that $r_1 < r_2$. Then, taking into account our hypotheses, we have

$$\left|(Vx^1, Vx^2)(r_2) - (Vx^1, Vx^2)(r_1)\right|$$
$$\leq |p(r_2) - p(r_1)|$$

$$
\begin{aligned}
&+\left|f(r_2,x^1(r_2),x^2(r_2))\int_0^{r_2} v(r_2,s,x^1(s),x^2(s))ds - f(r_1,x^1(r_1),x^2(r_1))\int_0^{r_2} v(r_2,s,x^1(s),x^2(s))ds\right|\\
&+\left|f(r_1,x^1(r_1),x^2(r_1))\int_0^{r_2} v(r_2,s,x^1(s),x^2(s))ds - f(r_1,x^1(r_1),x^2(r_1))\int_0^{r_1} v(r_1,s,x^1(s),x^2(s))ds\right|\\
&\leq \mu^{\mathbb{T}}(p,\epsilon)\\
&+\left|f(r_2,x^1(r_2),x^2(r_2)) - f(r_1,x^1(r_1),x^2(r_1))\right|\int_0^{r_2}\left[\left|v(r_2,s,x^1(s),x^2(s)) - v(r_2,s,0,0)\right| + |v(r_2,s,0,0)|\right]ds\\
&+\left|f(r_1,x^1(r_1),x^2(r_1))\right|\left|\int_0^{r_1} v(r_2,s,x^1(s),x^2(s))ds - \int_0^{r_1} v(r_1,s,x^1(s),x^2(s))ds + \int_{r_1}^{r_2} v(r_2,s,x^1(s),x^2(s))ds\right|\\
&\leq \mu^{\mathbb{T}}(p,\epsilon)\\
&+\left|f(r_2,x^1(r_2),x^2(r_2)) - f(r_1,x^1(r_1),x^2(r_1))\right|\left[\int_0^{r_2} h(r_2,s)H(\left|x^1(s)x^2(s)\right|)ds + \int_0^{r_2}|v(r_2,s,0,0)|\,ds\right]\\
&+\left|f(r_1,x^1(r_1),x^2(r_1))\right|\left[\int_0^{r_1}\left|v(r_2,s,x^1(s),x^2(s)) - v(r_1,s,x^1(s),x^2(s))\right|ds + \int_{r_1}^{r_2}\left|v(r_2,s,x^1(s),x^2(s))\right|ds\right]\\
&\leq \mu^{\mathbb{T}}(p,\epsilon)\\
&+\left[\left|f(r_2,x^1(r_2),x^2(r_2)) - f(r_2,x^1(r_1),x^2(r_1))\right|\right.\\
&\left.+\left|f(r_2,x^1(r_1),x^2(r_1)) - f(r_1,x^1(r_1),x^2(r_1))\right|\right]H(r_0)\int_0^{r_2} h(r_2,s)ds\\
&+\left[\left|f(r_2,x^1(r_2),x^2(r_2)) - f(r_2,x^1(r_1),x^2(r_1))\right|\right.\\
&\left.+\left|f(r_2,x^1(r_1),x^2(r_1)) - f(r_1,x^1(r_1),x^2(r_1))\right|\right]\int_0^{r_2}|v(r_2,s,0,0)|\,ds\\
&+\left[\left|f(r_1,x^1(r_1),x^2(r_1)) - f(r_1,0,0)\right| + |f(r_1,0,0)|\right]\left[\int_0^{r_1}\mu_1^{\mathbb{T}}(v,\epsilon,r_0)ds + \int_{r_1}^{r_2}\left|v(r_2,s,x^1(s),x^2(s))\right|ds\right]\\
&\leq \mu^{\mathbb{T}}(p,\epsilon)\\
&+\left[m(r_2)\left|x^1(r_2)x^2(r_2) - x^1(r_1)x^2(r_1)\right| + \mu_1^{\mathbb{T}}(f,\epsilon)\right]H(r_0)\int_0^{r_2} h(r_2,s)ds\\
&+\left[m(r_2)\left|x^1(r_2)x^2(r_2) - x^1(r_1)x^2(r_1)\right| + \mu_1^{\mathbb{T}}(f,\epsilon)\right]\int_0^{r_2}|v(r_2,s,0,0)|\,ds\\
&+\left[m(r_1)\left|x^1(r_1)x^2(r_1)\right| + |f(r_1,0,0)|\right]\left[\mathbb{T}\mu_1^{\mathbb{T}}(v,\epsilon,r_0) + \epsilon\bar{v}(\mathbb{T},r_0)\right]\\
&\leq \mu^{\mathbb{T}}(p,\epsilon)\\
&+m(r_2)\mu^{\mathbb{T}}((x^1,x^2),\epsilon)H(r_0)\int_0^{r_2} h(r_2,s)ds + \mu_1^{\mathbb{T}}(f,\epsilon)H(r_0)\sup\left\{\int_0^{\mathbb{T}} h(r,s)ds : r\in[0,\mathbb{T}]\right\}\\
&+m(r_2)\mu^{\mathbb{T}}((x^1,x^2),\epsilon)\int_0^{r_2}|v(r_2,s,0,0)|\,ds + \mu_1^{\mathbb{T}}(f,\epsilon)\sup\left\{\int_0^{\mathbb{T}}|v(r_2,s,0,0)|\,ds : r\in[0,\mathbb{T}]\right\}\\
&+[\bar{m}(\mathbb{T})r_0 + \bar{f}(\mathbb{T})][\mathbb{T}\mu_1^{\mathbb{T}}(v,\epsilon,r_0) + \epsilon\bar{v}(\mathbb{T},r_0)]
\end{aligned}
$$

$$
\begin{aligned}
&\leq \mu^{\mathbb{T}}(p,\epsilon)\\
&+[a(r_2)H(r_0) + c(r_2)]\mu^{\mathbb{T}}((x^1,x^2),\epsilon)\\
&+\mu_1^{\mathbb{T}}(f,\epsilon)H(r_0)\sup\left\{\int_0^{\mathbb{T}} h(r,s)ds : r\in[0,\mathbb{T}]\right\}\\
&+\mu_1^{\mathbb{T}}(f,\epsilon)\sup\left\{\int_0^{\mathbb{T}}|v(r_2,s,0,0)|\,ds : r\in[0,\mathbb{T}]\right\}\\
&+[\bar{m}(\mathbb{T})r_0 + \bar{f}(\mathbb{T})][\mathbb{T}\mu_1^{\mathbb{T}}(v,\epsilon,r_0) + \epsilon\bar{v}(\mathbb{T},r_0)]\\
&\leq \mu^{\mathbb{T}}(p,\epsilon)
\end{aligned}
$$

$$+ [AH(r_0) + C]\mu^{\mathbb{T}}((x^1, x^2), \epsilon)$$
$$+ \mu_1^{\mathbb{T}}(f, \epsilon)H(r_0) \sup\left\{\int_0^{\mathbb{T}} h(r, s)ds : r \in [0, \mathbb{T}]\right\}$$
$$+ \mu_1^{\mathbb{T}}(f, \epsilon) \sup\left\{\int_0^{\mathbb{T}} |v(r_2, s, 0, 0)|\, ds : r \in [0, \mathbb{T}]\right\}$$
$$+ [\bar{m}(\mathbb{T})r_0 + \bar{f}(\mathbb{T})][\mathbb{T}\mu_1^{\mathbb{T}}(v, \epsilon, r_0) + \epsilon\bar{v}(\mathbb{T}, r_0)].$$

where

$$\mu_1^{\mathbb{T}}(f, \epsilon) = \sup\left\{\begin{matrix} |f(r_2, x^1, x^2) - f(r_1, x^1, x^2)| : |r_2 - r_1| \leq \epsilon; r_1, r_2 \in [0, \mathbb{T}]; \\ x^1, x^2 \in [-r_0, r_0]. \end{matrix}\right\},$$

$$\bar{m}(\mathbb{T}) = \sup\left\{m(r) : r \in [0, \mathbb{T}]\right\},$$

$$\bar{f}(\mathbb{T}) = \sup\left\{|f(r, 0, 0)| : r \in [0, \mathbb{T}]\right\},$$

$$\mu_1^{\mathbb{T}}(v, \epsilon, r_0) = \sup\left\{\begin{matrix} |v(r_2, s, x^1, x^2) - v(r_1, s, x^1, x^2)| : |r_2 - r_1| \leq \epsilon; r_1, r_2, s \in [0, \mathbb{T}]; \\ s \leq r_1,\ s \leq r_2,\ x^1, x^2 \in [-r_0, r_0]. \end{matrix}\right\},$$

and

$$\bar{v}(\mathbb{T}, r_0) = \sup\left\{\begin{matrix} |v(r_2, s, x^1, x^2)| : r_1, r_2, s \in [0, \mathbb{T}]; \\ s \leq r_2,\ x^1, x^2 \in [-r_0, r_0]. \end{matrix}\right\}.$$

In addition, due to the uniform continuity of $f$ on $[0, \mathbb{T}] \times [-r_0, r_0]^2$ and $v$ on $[0, \mathbb{T}]^2 \times [-r_0, r_0]^2$, we deduce that $\mu^{\mathbb{T}}(p, \epsilon) \to 0$, $\mu_1^{\mathbb{T}}(f, \epsilon) \to 0$ and $\mu_1^{\mathbb{T}}(v, \epsilon, r_0) \to 0$ as $\epsilon \to 0$. we obtain

$$\mu_0^{\mathbb{T}}(V\mathscr{X}^1, V\mathscr{X}^2) \leq [AH(r_0) + C]\mu_0^{\mathbb{T}}(\mathscr{X}^1, \mathscr{X}^2).$$

i.e.

$$\mu_0^{\mathbb{T}}(V\mathscr{X}^1, V\mathscr{X}^2) \leq k\mu_0^{\mathbb{T}}(\mathscr{X}^1, \mathscr{X}^2).$$

where, $\mu$ is coupled MNC in $BC(\mathbb{R}_+)$.
Hereby Corollary 2.1, $\mathscr{D}$ has a fixed point in $B_{r_0}$.
i.e. the Eq. (4.1) has a solution in $BC(\mathbb{R}_+)$. □

**Illustration 4.1** Let us define the functional integral equation as follows

$$(x^1, x^2)(r) = re^{-2r} + \frac{1}{2\pi} tan^{-1}(\sqrt{r} + rx^1x^2(r)) \int_0^r \left[\frac{2(x^1x^2(r))^{\frac{2}{3}} + x^1x^2(r)}{(s+1)(r^2+1)} + \frac{1}{10(r^2+1)}\right]ds, \quad r \geq 0. \tag{4.2}$$

this equation is a special case of Eq. (4.1).

Here

$$p(r) = re^{-2r},$$

$$f(r, x^1(r), x^2(r)) = \frac{1}{2\pi} tan^{-1}(\sqrt{r}, x^1x^2(r)),$$

and

$$v(r, s, x^1(s), x^2(s)) = \frac{2(x^1x^2(r))^{\frac{2}{3}} + x^1x^2(r)}{(s+1)(r^2+1)} + \frac{1}{10(r^2+1)}.$$

Then, we can check that the hypotheses of Theorem (4.1) are satisfied.
Indeed, we have that

$$\| p \| = p(1/2) = \frac{1}{2}e^{-1} = 0.18394,$$

$$f(r, 0, 0) = \frac{1}{2\pi} tan^{-1}(\sqrt{r}),$$

and

$$v(r, s, 0, 0) = \frac{1}{10(r^2+1)}.$$

Further, we have that

$$m(r) = \frac{1}{2\pi},$$

$$\begin{aligned}\left|v(r, s, x^1, x^2) - v(r, s, x_1^1, x_1^2)\right| &\le \frac{1}{(s+1)(r^2+1)} \left|2(x^1x^2)^{\frac{2}{3}} + x^1x^2 - 2(x_1^1x_1^2)^{\frac{2}{3}} - x_1^1x_1^2\right| \\ &\le \frac{1}{(s+1)(r^2+1)}[2((x^1x^2)^{\frac{2}{3}} - (x_1^1x_1^2)^{\frac{2}{3}}) + \left|x^1x^2 - x_1^1x_1^2\right|].\end{aligned}$$

So, the functions appearing in hypothesis $(H3)$ have the form:

$$h(r, s) = \frac{1}{(s+1)(r^2+1)}$$

and

$$H(r) = 2(r)^{\frac{2}{3}} + r.$$

Further, we deduce that the function $a(r)$ appearing in hypothesis $(H4)$ has the following form:

$$a(r) = m(r) \int_0^r h(r, s)ds = \frac{1}{2\pi} \frac{r \ln(r+1)}{r^2+1},$$

So, we have that $a(r) \to 0$ as $r \to \infty$. Apart from this, using the estimate $\ln(r+1) \leq \sqrt{r}$, we obtain

$$a(r) \leq \frac{1}{2\pi}\frac{r\sqrt{r}}{r^2+1},$$

$$A \leq \frac{1}{2\pi}\frac{3^{3/4}}{4} = 0.09069.$$

Next, we obtain

$$\begin{aligned} b(r) &= |f(r,0,0)| \int_0^r h(r,s)ds \\ &= \frac{1}{2\pi} tan^{-1}(\sqrt{r}) \frac{r\ln(r+1)}{r^2+1} \leq \frac{1}{4}\frac{\sqrt{r}}{r^2+1}. \end{aligned}$$

Thus, $b(r) \to 0$ as $r \to \infty$. Moreover, we have that

$$B \leq \frac{3}{16}\frac{1}{\sqrt[4]{3}} = 0.14247.$$

Further,

$$\begin{aligned} c(r) &= m(r) \int_0^r |v(r,s,0,0)|\, ds \\ &= \frac{1}{2\pi}\int_0^r \frac{1}{10(r^2+1)} ds = \frac{1}{20\pi}\frac{r^2}{r^2+1} ds. \end{aligned}$$

Hence, $C = 0.01592$.
Finally, we obtain

$$\begin{aligned} d(r) &= |f(r,0,0)| \int_0^r |v(r,s,0,0)|\, ds \\ &= \frac{1}{2\pi} tan^{-1}(\sqrt{r}) \frac{r}{10(r^2+1)} \leq \frac{1}{40}\frac{r}{r^2+1}. \end{aligned}$$

This gives that $D \leq 1/80 = 0.0125$.

In order to verify hypothesis $(H5)$, observe that the inequality appearing in this hypothesis has the form:

$$\| p \| + A(2r_0^{5/3} + r_0^2) + B(2r_0^{5/3} + r_0^2) + Cr_0 + D \leq r_0.$$

it is easy to check that the number $r_0 = 1$ satisfies the above inequality. Moreover, we have that $k = AH(r_0) + C = 3A + C \leq 0.288012 < 1$. Hence, hypothesis $(H5)$ is satisfied.

Thus, we have achieved all of the assumptions from $(H1)$ to $(H5)$ in Theorem 4.1.
From Theorem 4.1, we can say that the Eq. (4.2) have solutions in $BC(\mathbb{R}_+)$.

## References


1. J. Banaś, K. Goebel, *Measure of Noncompactness in Banach Spaces*, Lecture Notes in Pure and Applied Mathematics, vol. 60 (Marcel Dekker, New York, 1980)
2. R.P. Agarwal, M. Meehan, D. O'regan, *Fixed Point Theory and Applications* (Cambridge University Press, 2001)
3. G. Darbo, Punti uniti in trasformazioni a codominio non compatto. Rendiconti del Seminario matematico della Università di Padova **24**, 84–92 (1955)
4. H. Hosseinzadeh, H. Işık, S.H. Bonab, R. George, Coupled measure of noncompactness and functional integral equations. Open Math. **20**(1), 38–49 (2022)
5. M.S. Khan, M. Swaleh, S. Sessa, Fixed point theorems by altering distances between the points. Bull. Australian Math. Soc. **30**(1), 1–9 (1984)
6. A. Aghajani, J. Banaś, N. Sabzali, Some generalizations of Darbo fixed point theorem and applications. Bull. Belgian Math. Soc.-Simon Stevin **20**(2), 345–358 (2013)
7. A.H. Ansari, Note on $\phi - \psi$ contractive type mappings and related fixed point, in *The 2nd Regional Conference on Mathematics and Applications* (Payame Noor University, 2014), pp. 377–380
8. I. Haque, J. Ali, M. Mursaleen, Solvability of implicit fractional order integral equation in $l_p(1 \leq p<\infty)$ space via generalized Darbo's fixed point theorem. J. Funct. Spaces (2022)
9. R. Arab, H.K. Nashine, N.H. Can, T.T. Binh, Solvability of functional-integral equations (fractional order) using measure of noncompactness. Adv. Diff. Equ. **2020**(1), 1–13 (2020)
10. A. Aghajani, M. Aliaskari, Measure of noncompactness in Banach algebra and application to the solvability of integral equations in $BC(\mathbb{R}_+)$, Inf. Sci. Lett. **4**(2), 93–99 (2015)
11. N. Ahmad, A. Al-Rawashdeh, N. Mehmood, S. Radenović, Fixed points of monotone mappings via generalized-measure of noncompactness. Vietnam J. Math. **50**(1), 275–285 (2022)
12. J. Appell, Measures of noncompactness, condensing operators and fixed points: an application-oriented survey. Fixed Point Theory **6**(2), 157–229 (2005)
13. S. Chang, Y. Cho, N. Huang, Coupled fixed point theorems with applications. J. Korean Math. Soc. **33**(3), 575–585 (1996)
14. S. Beloul, M. Mursaleen, A.H. Ansari, A generalization of Darbo's fixed point theorem with an application to fractional integral equations. J. Math. Ineq. **15**, 911–921 (2021)
15. A. Das, S.A. Mohiuddine, A. Alotaibi, B.C. Deuri, Generalization of Darbo-type theorem and application on existence of implicit fractional integral equations in tempered sequence spaces. Alexandria Eng. J. **61**(3), 2010–2015 (2022)
16. J. Banaś, Measures of noncompactness in the study of solutions of nonlinear differential and integral equations. Open Math. **10**(6), 2003–2011 (2012)
17. A. Das, B. Hazarika, R. Arab, M. Mursaleen, Applications of a fixed point theorem to the existence of solutions to the nonlinear functional integral equations in two variables. Rendiconti del Circolo Matematico di Palermo Ser. **2**, 68(1), 139–152 (2019)
18. J. Banaś, On measures of noncompactness in Banach spaces. Commentationes Mathematicae Universitatis Carolinae **21**(1), 131–143 (1980)
19. A. Das, B. Hazarika, V. Parvaneh, M. Mursaleen, Solvability of generalized fractional order integral equations via measures of noncompactness. Math. Sci. **15**(3), 241–251 (2021)
20. K. Kuratowski, Sur les espaces complets. Fundamenta Mathematicae **1**(15), 301–309 (1930)
21. B. Hazarika, R. Arab, M. Mursaleen, Applications of measure of noncompactness and operator type contraction for existence of solution of functional integral equations. Compl. Anal. Oper. Theory **13**(8), 3837–3851 (2019)

22. B. Hazarika, R. Arab, M. Mursaleen, Application measure of noncompactness and operator type contraction for solvability of an infinite system of differential equations in lp-space. Filomat **33**(7), 2181–2189 (2019)
23. A. Das, V. Parvaneh, B.C. Deuri, Z. Bagheri, Application of a generalization of Darbo's fixed point theorem via Mizogochi-Takahashi mappings on mixed fractional integral equations involving $(k, s)$-Riemann-Liouville and Erd'elyi-Kober fractional integrals. Int. J. Nonlinear Anal. Appl. **13**(1), 859–869 (2022)
24. V. Parvaneh, N. Hussain, Z. Kadelburg, Generalized Wardowski type fixed point theorems via $\alpha$-admissible FG-contractions in b-metric spaces. Acta Mathematica Scientia **36**(5), 1445–1456 (2016)
25. M. Mursaleen, Differential equations in classical sequence spaces. Revista de la Real Academia de Ciencias Exactas, Físicas y Naturales. Serie A. Matemáticas **111**(2), 587–612 (2017)
26. H.K. Nashine, R. Arab, R.P. Agarwal, A.S. Haghighi, Darbo type fixed and coupled fixed point results and its application to integral equation. Periodica Mathematica Hungarica **77**(1), 94–107 (2018)
27. M. Rabbani, A. Das, B. Hazarika, R. Arab, Measure of noncompactness of a new space of tempered sequences and its application on fractional differential equations. Chaos Solit. Fract. **140**, 110221 (2020)
28. M. Gabeleh, C. Vetro, A new extension of Darbo's fixed point theorem using relatively Meir-Keeler condensing operators. Bull. Australian Math. Soc. **98**(2), 286–297 (2018)
29. H.K. Nashine, R. Arab, R.P. Agarwal, M. De la Sen, Positive solutions of fractional integral equations by the technique of measure of noncompactness. J. Inequal. Appl. **2017**(1), 1–16 (2017)
30. B.C. Deuri, Some new fixed point theorem via shifting distance functions. Appl. Appl. Math.: An Int. J. (AAM) **17**(3), 6 (2022)
31. S.A. Mohiuddine, A. Das, A. Alotaibi, Existence of solutions for nonlinear integral equations in tempered sequence spaces via generalized Darbo-Type theorem. J. Funct. Spaces (2022)
32. I. Haque, J. Ali, M. Mursaleen, Solvability of implicit fractional order integral equation in $\ell_p$ $(1<p<\infty)$ space via generalized Darbo's fixed point theorem. J. Funct. Spaces (2022)
33. B.C. Deuri, A. Das, Solvability of fractional integral equations via Darbo's fixed point theorem. J. Pseudo-Diff. Oper. Appl. **13**(3), 1–12 (2022)
34. H.K. Nashine, A. Das, Extension of Darbo's fixed point theorem via shifting distance functions and its application. Nonlinear Anal.: Modell. Control **27**(2), 275–288 (2022)

# Generalized Bernstein Operators Associated with Multiple Shape Parameters

**Uğur Kadak and Faruk Özger**

**Abstract** In this chapter, a novel extension of Bernstein type operators are introduced based on multiple shape parameters $\lambda_i, i = 1, 2, \ldots, n$. Convergence properties of the operators are investigated and local and global approximation theorems are established via Ditzian-Totik modulus of smoothness of first and second order. The rate of convergence is estimated with the help of Lipschitz-type function involving two parameters. Furthermore, statistical approximation properties are given using weighted mean matrix method, and Voronovskaja type asymptotic approximation results are presented. Finally, some illustrative graphs and numerical examples are provided to show the efficiency, applicability and validity of proposed operators.

**Keywords** Bernstein operators · Modulus of continuity · Rate of convergence · Lipschitz continuous functions · Voronovskaja asymptotic formula

**2010 Mathematics Subject Classification** Primary 41A10 · 41A25 · 41A36 · Secondary 26A16 · 40C05

## 1 Introduction

Approximation of real and complex-valued continuous functions defined on a compact interval or whole real line has recently attracted attention of many researchers from different disciplines such as mathematics, physics, engineering, economics and computational sciences. Classical approximation theorem of Weierstrass (what is subsequently known as his approximation theorem) plays a fundamental role in development of approximation theory (see [1]). This states that, for any continuous

U. Kadak
Department of Mathematics, Gazi University, Ankara, Türkiye

Yaşamkent Street, Bolu, Türkiye

F. Özger (✉)
Department of Computer Engineering, Iğdır University, Iğdır, Türkiye
e-mail: farukozger@gmail.com

 
S. A. Mohiuddine et al. (eds.), *Positive Operators and Fixed-Point Theorems with Applications*, Industrial and Applied Mathematics,
https://doi.org/10.1007/978-981-96-2476-8_6

function $f(x)$ defined on a closed interval $[a, b]$ and for each $\epsilon > 0$, there is an algebraic polynomial $p_n(x)$ of sufficiently high degree $n$ such that $|f(x) - p_n(x)| < \epsilon$ for all $x \in [a, b]$. To obtain an alternative proof of Weierstrass's fundamental theorem, S. N. Bernstein [2] proposed famous polynomials nowadays called Bernstein polynomials, in 1912. Bernstein basis of degree $n$ on $x \in [0, 1]$ is defined by

$$b_{n,i}(x) = \binom{n}{i} x^i (1-x)^{n-i}, \qquad i = 0, \dots, n,$$

and $n$th order Bernstein polynomial is specified as

$$B_n(f; x) = \sum_{i=0}^{n} f\left(\frac{i}{n}\right) b_{n,i}(x) \tag{1.1}$$

for any continuous function $f$ defined on [0, 1]. For more details about the Bernstein polynomials, see, e.g., [3–5].

In recent years, efforts of many mathematicians have been focused on generalizations of the classical positive linear operators, see [6–13], and they have numerous applications including dynamic programming, the binomial theorem, probability, discrete convolution, integral equations, Computer Aided Geometric Design (CAGD) and Computer Graphics (CG); see [14–20]. In 2008, Han et al. [21] introduced a new class of Bernstein basis functions with adjustable shape parameters as follows.

For $n$ $(n \geq 2)$ arbitrary real values of $\lambda_i, i = 1, 2, \dots, n$, the following polynomial functions in $x \in [0, 1]$

$$\begin{cases} a_{n,0}(\lambda; x) = (1-x)^n (1 - \lambda_1 x), \\ a_{n,i}(\lambda; x) = x^i (1-x)^{n-i} \left(\binom{n}{i} + \lambda_i - \lambda_i x - \lambda_{i+1} x\right), \; i = 1, 2, \dots, [\frac{n}{2}] - 1, \\ a_{n,[\frac{n}{2}]}(\lambda; x) = x^{[\frac{n}{2}]}(1-x)^{n-[\frac{n}{2}]}\left(\binom{n}{[\frac{n}{2}]} + \lambda_{[\frac{n}{2}]} - \lambda_{[\frac{n}{2}]} x + \lambda_{[\frac{n}{2}]+1} x\right), \\ a_{n,i}(\lambda; x) = x^i (1-x)^{n-i} \left(\binom{n}{i} - \lambda_i + \lambda_i x + \lambda_{i+1} x\right), \; i = [\frac{n}{2}] + 1, \dots, n-1, \\ a_{n,n}(\lambda; x) = x^n (1 - \lambda_n + \lambda_n x) \end{cases} \tag{1.2}$$

are called generalized Bernstein polynomials of degree $n$ with shape parameters $\lambda_i$, $i = 1, 2, \dots, n$, where

$$\begin{cases} \lambda_i \in [-\binom{n}{i}, \binom{n}{i-1}] & ; i = 1, 2, \dots, [\frac{n}{2}] \\ \lambda_i \in [-\binom{n}{i-1}, \binom{n}{i}] & ; i = [\frac{n}{2}] + 1, \dots, n \end{cases} \quad \text{with} \quad \begin{cases} [\frac{n}{2}] = \frac{n}{2} & ; \text{if } n \text{ is even} \\ [\frac{n}{2}] = \frac{n-1}{2} & ; \text{if } n \text{ is odd.} \end{cases} \tag{1.3}$$

In particular, for $\lambda_i = 0$ $(i = 1, 2, \dots, n)$, the generalized Bernstein basis functions defined by (1.2) are reduced to classical Bernstein basis functions in (1.1). It can

be easily proved that generalized Bernstein basis functions with parameters $\lambda_i$ $(i = 1, 2, \ldots, n)$ are linearly independent (see [14]). Previous works have showed that these basis functions are effective and flexible, and play a significant role designing parametric curves and surfaces (i.e., CAGD) (see [14, 21]). The generalized Bernstein basis functions (1.2) have the following properties (see [21]):
(1) partition of unity: $\sum_{i=0}^{n} a_{n,i}(\lambda; x) \equiv 1$.
(2) nonnegativity: $a_{n,i}(\lambda; x) \geq 0$ $(i = 0, 1, \ldots, n)$.
(3) symmetry: for $i = 0, 1, \ldots, n$;

$$a_{n,i}(\lambda_1, \ldots, \lambda_{[\frac{n}{2}]}, \lambda_{[\frac{n}{2}]+1}, \ldots, \lambda_n; x) = a_{n,n-i}(\lambda_n, \ldots, \lambda_{[\frac{n}{2}]+1}, (-1)^n\lambda_{[\frac{n}{2}]}, \ldots, \lambda_1; 1-x).$$

In this paper, a novel extension of classical Bernstein type operator is constructed based on the ideas exposed in [21]. More precisely, we aim to introduce a new kind of Bernstein type operators involving multiple shape parameters for the first time.

## 2 Construction of Operators

Let $C[0, 1]$ be the space of all continuous functions on unit interval $[0, 1]$. For $n$ arbitrary real values of $\lambda_i$ $(i = 0, 1, \ldots n)$ and $x \in [0, 1]$, the following operators

$$B_n(f; x; \lambda) = \sum_{i=0}^{n} f\left(\frac{i}{n}\right) a_{n,i}(\lambda; x) \tag{2.1}$$

are called generalized Bernstein operators associated with the shape parameters $\lambda_i$ satisfying (1.3). In 2017, Hu et al. [14] have obtained the following equations to convert classical Bernstein polynomials of degree $n$ to generalized Bernstein polynomials of degree $n + 1$ associated with shape parameters $\lambda_i$:

$$\begin{cases} a_{n,0}(\lambda; x) = b_{n+1,0}(x) + \frac{\binom{n}{0}-\lambda_1}{\binom{n+1}{1}} b_{n+1,1}(x), \\ a_{n,i}(\lambda; x) = \frac{\binom{n}{i}+\lambda_i}{\binom{n+1}{i}} b_{n+1,i}(x) + \frac{\binom{n}{i}-\lambda_{i+1}}{\binom{n+1}{i+1}} b_{n+1,i+1}(x), \quad i = 1, 2 \ldots, \left[\frac{n}{2}\right] - 1, \\ a_{n,i}(\lambda; x) = \frac{\binom{n}{i}+\lambda_i}{\binom{n+1}{i}} b_{n+1,i}(x) + \frac{\binom{n}{i}+\lambda_{i+1}}{\binom{n+1}{i+1}} b_{n+1,i+1}(x), \ i = \left[\frac{n}{2}\right], \\ a_{n,i}(\lambda; x) = \frac{\binom{n}{i}-\lambda_i}{\binom{n+1}{i}} b_{n+1,i}(x) + \frac{\binom{n}{i}+\lambda_{i+1}}{\binom{n+1}{i+1}} b_{n+1,i+1}(x), \quad i = \left[\frac{n}{2}\right] + 1, \ldots, n-1, \\ a_{n,n}(\lambda; x) = \frac{\binom{n}{n}-\lambda_n}{\binom{n+1}{n}} b_{n+1,n}(x) + b_{n+1,n+1}(x), \end{cases} \tag{2.2}$$

where $b_{n+1,i}(x) = \binom{n+1}{i} x^i (1-x)^{n+1-i}, \quad x \in [0, 1], \ i = 0, \ldots, n+1$, are Bernstein basis functions of degree $n + 1$.

**Lemma 2.1** *For any natural number n ($n \geq 2$), the following equalities hold:*

$$B_n(1;x;\lambda)=1,\ B_n(t;x;\lambda)=x+\frac{1-x}{n}\mathcal{R}_0(x),\ B_n(t^2;x;\lambda)=\frac{n-1}{n}x^2+\frac{x}{n}+\frac{1-x}{n^2}[2\mathcal{R}_1(x)-\mathcal{R}_0(x)],$$

$$B_n(t^3;x;\lambda)=\frac{(n-1)(n-2)}{n^2}x^3+\frac{3(n-1)}{n^2}x^2+\frac{x}{n^2}+\frac{1-x}{n^3}[3\mathcal{R}_2(x)-3\mathcal{R}_1(x)+\mathcal{R}_0(x)], \tag{2.3}$$

$$B_n(t^4;x;\lambda)=\frac{(n-1)(n-2)(n-3)}{n^3}x^4+\frac{6(n-1)(n-2)}{n^3}x^3+\frac{7(n-1)}{n^3}x^2+\frac{x}{n^3}$$
$$+\frac{1-x}{n^4}[4\mathcal{R}_3(x)-6\mathcal{R}_2(x)+4\mathcal{R}_1(x)-\mathcal{R}_0(x)], \tag{2.4}$$

*where*

$$\mathcal{R}_p(x)=\sum_{i=1}^{[\frac{n}{2}]}\frac{i^p\lambda_i}{\binom{n}{i}}b_{n,i}(x)-\sum_{i=[\frac{n}{2}]+1}^{n}\frac{i^p\lambda_i}{\binom{n}{i}}b_{n,i}(x),\ p=0,1,2,\ldots n,\ x\in[0,1]. \tag{2.5}$$

***Proof*** Let $\lambda_i$ ($i=1,\ldots n$) be arbitrary real shape parameters satisfying (1.3) and $x \in [0,1]$. By partition of unity property, the generalized Bernstein operators $B_n(e_0;x;\lambda)$ reproduce the constant function $e_0(x)=1$. Let $n=2m, m\in\mathbb{N}$, and $x\in[0,1]$. Using (2.2) we write that

$$\begin{aligned}
&B_{2m}(t;x;\lambda)\\
&=\sum_{i=0}^{m-1}\frac{i}{2m}\frac{\binom{2m}{i}+\lambda_i}{\binom{2m+1}{i}}b_{2m+1,i}(x)+\sum_{i=0}^{m-1}\frac{i}{2m}\frac{\binom{2m}{i}-\lambda_{i+1}}{\binom{2m+1}{i+1}}b_{2m+1,i+1}(x)+\sum_{i=m+1}^{2m-1}\frac{i}{2m}\frac{\binom{2m}{i}-\lambda_i}{\binom{2m+1}{i}}b_{2m+1,i}(x)\\
&+\sum_{i=m+1}^{2m-1}\frac{i}{2m}\frac{\binom{2m}{i}+\lambda_{i+1}}{\binom{2m+1}{i+1}}b_{2m+1,i+1}(x)+\frac{1}{2}\left\{\frac{\binom{2m}{m}+\lambda_m}{\binom{2m+1}{m}}b_{2m+1,m}(x)\right\}+\frac{1}{2}\left\{\frac{\binom{2m}{m}+\lambda_{m+1}}{\binom{2m+1}{m+1}}b_{2m+1,m+1}(x)\right\}\\
&+\frac{\binom{2m}{2m}-\lambda_{2m}}{\binom{2m+1}{2m}}b_{2m+1,2m}(x)+b_{2m+1,2m+1}(x)\\
&=\sum_{i=0}^{2m-1}\frac{i}{2m}\binom{2m}{i}x^i(1-x)^{2m-i+1}+\sum_{i=0}^{2m-1}\frac{i}{2m}\binom{2m}{i}x^{i+1}(1-x)^{2m-i}+\sum_{i=0}^{m-1}\frac{i}{2m}\lambda_i\,x^i(1-x)^{2m-i+1}\\
&-\sum_{i=m+1}^{2m-1}\frac{i}{2m}\lambda_i\,x^i(1-x)^{2m-i+1}-\sum_{i=0}^{m-1}\frac{i}{2m}\lambda_{i+1}x^{i+1}(1-x)^{2m-i}+\sum_{i=m+1}^{2m-1}\frac{i}{2m}\lambda_{i+1}x^{i+1}(1-x)^{2m-i}\\
&+\frac{1}{2}\left\{\lambda_m x^m(1-x)^{m+1}\right\}+\frac{1}{2}\left\{\lambda_{m+1}x^{m+1}(1-x)^m\right\}+x^{2m}(1-x)-\lambda_{2m}x^{2m}(1-x)+x^{2m+1}\\
&=(1-x)\sum_{i=0}^{2m}\frac{i}{2m}\binom{2m}{i}x^i(1-x)^{2m-i}+x\sum_{i=0}^{2m}\frac{i}{2m}\binom{2m}{i}x^i(1-x)^{2m-i}+U_1+U_2+U_3+U_4
\end{aligned}$$

where

$$U_1 = \sum_{i=1}^{m} \frac{i}{2m}\lambda_i\, x^i(1-x)^{2m-i+1} - \sum_{i=0}^{m-1} \frac{i}{2m}\lambda_{i+1}x^{i+1}(1-x)^{2m-i},$$

$$U_2 = -\sum_{i=m+1}^{2m-1} \frac{i}{2m}\,\lambda_i\, x^i(1-x)^{2m-i+1} + \sum_{i=m+1}^{2m-1} \frac{i}{2m}\lambda_{i+1}x^{i+1}(1-x)^{2m-i},$$

$$U_3 = \frac{1}{2}\left\{\lambda_m x^m(1-x)^{m+1}\right\} + \frac{1}{2}\left\{\lambda_{m+1}x^{m+1}(1-x)^m\right\} - \lambda_{2m}x^{2m}(1-x).$$

Since

$$\begin{aligned} U_1 &= \sum_{i=1}^{m} \frac{i}{2m}x^i(1-x)^{2m-i}[\lambda_i(1-x) - \lambda_{i+1}x] = \frac{1}{2m}x(1-x)^{2m}\lambda_1 - \frac{1}{2m}\,x^2(1-x)^{2m-1}\lambda_2 \\ &+ \frac{2}{2m}\,x^2(1-x)^{2m-1}\lambda_2 - \frac{2}{2m}\,x^3(1-x)^{2m-2}\lambda_3 + \frac{3}{2m}\,x^3(1-x)^{2m-2}\lambda_3 - \frac{3}{2m}\,x^4(1-x)^{2m-3}\lambda_4 + \cdots \\ &+ \frac{m-1}{2m}\,x^{m-1}(1-x)^{m+2}\lambda_{m-1} - \frac{m-1}{2m}\,x^m(1-x)^{m+1}\lambda_m + \frac{m}{2m}\,x^m(1-x)^{m+1}\lambda_m - \frac{m}{2m}\,x^{m+1}(1-x)^m\lambda_{m+1} \\ &= \frac{1}{2m}x(1-x)^{2m}\lambda_1 + \frac{1}{2m}\,x^2(1-x)^{2m-1}\lambda_2 + \frac{1}{2m}\,x^3(1-x)^{2m-2}\lambda_3 + \frac{1}{2m}\,x^4(1-x)^{2m-3}\lambda_4 + \cdots \\ &+ \frac{1}{2m}\,x^m(1-x)^{m+1}\lambda_m - \frac{1}{2}\,x^{m+1}(1-x)^m\lambda_{m+1} \end{aligned}$$

and

$$\begin{aligned} U_2 &= \sum_{i=m}^{2m-1} \frac{i}{2m}x^i(1-x)^{2m-i}[\lambda_{i+1}x - \lambda_i(1-x)] = \frac{m}{2m}x^{m+1}(1-x)^m\lambda_{m+1} - \frac{m}{2m}x^m(1-x)^{m+1}\lambda_m \\ &+ \frac{m+1}{2m}x^{m+2}(1-x)^{m-1}\lambda_{m+2} - \frac{m+1}{2m}x^{m+1}(1-x)^m\lambda_{m+1} + \frac{m+2}{2m}x^{m+3}(1-x)^{m-2}\lambda_{m+3} \\ &- \frac{m+2}{2m}x^{m+2}(1-x)^{m-1}\lambda_{m+2} + \cdots + \frac{2m-2}{2m}x^{2m-1}(1-x)^2\lambda_{2m-1} - \frac{2m-2}{2m}x^{2m-2}(1-x)^3\lambda_{2m-2} \\ &+ \frac{2m-1}{2m}x^{2m}(1-x)\lambda_{2m} - \frac{2m-1}{2m}x^{2m-1}(1-x)^2\lambda_{2m-1} \\ &= -\frac{1}{2m}\sum_{i=m+1}^{2m-1} x^i(1-x)^{2m-i+1}\lambda_i - \frac{1}{2}x^m(1-x)^{m+1}\lambda_m + \frac{2m-1}{2m}x^{2m}(1-x)\lambda_{2m}, \end{aligned}$$

we get

$$U_1 + U_2 + U_3 = \frac{1-x}{2m}\sum_{i=1}^{m} x^i(1-x)^{2m-i}\lambda_i - \frac{1-x}{2m}\sum_{i=m+1}^{2m} x^i(1-x)^{2m-i}\lambda_i.$$

By using the last equality, we observe that

$$\begin{aligned} B_{2m}(t;x;\lambda) &= B_{2m,i}(t;x) + \frac{1-x}{2m}\sum_{i=1}^{m} x^i(1-x)^{2m-i}\lambda_i - \frac{1-x}{2m}\sum_{i=m+1}^{2m} x^i(1-x)^{2m-i}\lambda_i \\ &= x + \frac{1-x}{2m}\mathcal{R}_0(x), \end{aligned}$$

which yields $B_n(t; x; \lambda) = x + \dfrac{1-x}{n}\mathcal{R}_0(x)$. We omit the computations for $n$ being an odd number. Similarly, for $n = 2m$, we may calculate $B_n(t^2; x; \lambda)$ as

$$
\begin{aligned}
B(t^2; x; \lambda) &= \sum_{i=1}^{m-1}\left(\frac{i}{2m}\right)^2 \frac{\binom{2m}{i}+\lambda_i}{\binom{2m+1}{i}} b_{2m+1,i}(x) + \sum_{i=1}^{m-1}\left(\frac{i}{2m}\right)^2 \frac{\binom{2m}{i}-\lambda_{i+1}}{\binom{2m+1}{i+1}} b_{2m+1,i+1}(x) \\
&+ \sum_{i=m+1}^{2m-1}\left(\frac{i}{2m}\right)^2 \frac{\binom{2m}{i}-\lambda_i}{\binom{2m+1}{i}} b_{2m+1,i}(x) + \sum_{i=m+1}^{2m-1}\left(\frac{i}{2m}\right)^2 \frac{\binom{2m}{i}+\lambda_{i+1}}{\binom{2m+1}{i+1}} b_{2m+1,i+1}(x) \\
&+ \frac{1}{4}\left\{\frac{\binom{2m}{m}+\lambda_m}{\binom{2m+1}{m}} b_{2m+1,m}(x)\right\} + \frac{1}{4}\left\{\frac{\binom{2m}{m}+\lambda_{m+1}}{\binom{2m+1}{m+1}} b_{2m+1,m+1}(x)\right\} \\
&+ \frac{\binom{2m}{2m}-\lambda_{2m}}{\binom{2m+1}{2m}} b_{2m+1,2m}(x) + b_{2m+1,2m+1}(x) \\
&= (1-x)\sum_{i=1}^{2m}\left(\frac{i}{2m}\right)^2\binom{2m}{i}x^i(1-x)^{2m-i} + x\sum_{i=1}^{2m}\left(\frac{i}{2m}\right)^2\binom{2m}{i}x^i(1-x)^{2m-i} \\
&+ V_1 + V_2 + V_3
\end{aligned}
$$

where

$$
\begin{aligned}
V_1 &= \sum_{i=1}^{m-1}\left(\frac{i}{2m}\right)^2 \lambda_i\, x^i(1-x)^{2m-i+1} - \sum_{i=1}^{m-1}\left(\frac{i}{2m}\right)^2 \lambda_{i+1}x^{i+1}(1-x)^{2m-i}, \\
V_2 &= -\sum_{i=m+1}^{2m-1}\left(\frac{i}{2m}\right)^2 \lambda_i\, x^i(1-x)^{2m-i+1} + \sum_{i=m+1}^{2m-1}\left(\frac{i}{2m}\right)^2 \lambda_{i+1}x^{i+1}(1-x)^{2m-i}, \\
V_3 &= \frac{1}{4}\left\{\lambda_m x^m(1-x)^{m+1}\right\} + \frac{1}{4}\left\{\lambda_{m+1}x^{m+1}(1-x)^m\right\} - \lambda_{2m}x^{2m}(1-x).
\end{aligned}
$$

Since

$$
V_1 = \frac{1}{4m^2}\sum_{i=1}^{m} x^i(1-x)^{2m-i+1}[i^2-(i-1)^2)]\lambda_i - \frac{1}{4}\,x^{m+1}(1-x)^m\lambda_{m+1}
$$

and

$$
V_2 = \frac{1}{4m^2}\sum_{i=m+1}^{2m-1} x^i(1-x)^{2m-i+1}[(i-1)^2-i^2]\lambda_i - \frac{1}{4}x^m(1-x)^{m+1}\lambda_m + \left(\frac{2m-1}{2m}\right)^2 x^{2m}(1-x)\lambda_{2m},
$$

we get

$$
\begin{aligned}
&V_1 + V_2 + V_3 \\
&= \frac{1}{4m^2}\sum_{i=1}^{m} x^i(1-x)^{2m-i+1}[i^2-(i-1)^2)]\lambda_i - \frac{1}{4}x^{m+1}(1-x)^m\lambda_{m+1} \\
&+ \frac{1}{4m^2}\sum_{i=m+1}^{2m-1} x^i(1-x)^{2m-i+1}[(i-1)^2-i^2]\lambda_i - \frac{1}{4}x^m(1-x)^{m+1}\lambda_m + \left(\frac{2m-1}{2m}\right)^2 x^{2m}(1-x)\lambda_{2m} \\
&+ \frac{1}{4}\left\{\lambda_m x^m(1-x)^{m+1}\right\} + \frac{1}{4}\left\{\lambda_{m+1}x^{m+1}(1-x)^m\right\} - \lambda_{2m}x^{2m}(1-x).
\end{aligned}
$$

Therefore

$$
\begin{aligned}
B_{2m}(t^2;x;\lambda) &= B_{2m,i}(t^2;x) + \frac{1-x}{4m^2}\sum_{i=1}^{m} x^i(1-x)^{2m-i}(2i-1)\lambda_i - \frac{1-x}{4m^2}\sum_{i=m+1}^{2m} x^i(1-x)^{2m-i}(2i-1)\lambda_i \\
&= x^2 + \frac{x(1-x)}{2m} + \frac{1-x}{4m^2}[2\mathcal{R}_1(x) - \mathcal{R}_0(x)].
\end{aligned}
$$

For $n = 2m$, we have

$$
B_n(t^2;x;\lambda) = x^2 + \frac{x(1-x)}{n} + \frac{1-x}{n^2}\big[2\mathcal{R}_1(x) - \mathcal{R}_0(x)\big].
$$

Obviously, when $n$ is taken as an odd number, the proof is obtained in a similar way. For the sake of brevity, we omit proofs of (2.3) and (2.4) since they are similar to the above proofs. □

**Remark 2.1** In view of Lemma 2.1, one can see that $B_n(e_i;x;\lambda) \to e_i(x)$ as $n \to \infty$, for all $i \in \mathbb{N}$ and $e_i(x) = x^i$, $i \in \mathbb{N}$. That is to say that the generalized Bernstein type operators $B_n(f;x;\lambda)$ reproduce linear polynomials in the limit case.

**Corollary 2.1** *For $n$ arbitrary real values of $\lambda_i$, where $n \geq 2$, $i = 1, 2, \ldots, n$, the central moments of generalized Bernstein polynomials with shape parameters $\lambda_i$ introduced in (2.1) are given below*

$$
B_n(t-x;x;\lambda) = \frac{1-x}{n}\mathcal{R}_0(x); \tag{2.6}
$$

$$
B_n((t-x)^2;x;\lambda) = \frac{x(1-x)}{n}(1-2\mathcal{R}_0(x)) + \frac{1-x}{n^2}\big[2\mathcal{R}_1(x) - \mathcal{R}_0(x)\big] \tag{2.7}
$$

*where $\mathcal{R}_p(x)$, $p = 0, 1, 2, \ldots n$, is defined as in Lemma 2.1.*

**Theorem 2.1** *For fixed $x \in [0, 1]$ and $n$ arbitrary real values of $\lambda_i$ ($n \geq 2$) satisfying (1.3), we have*

$$
\lim_{n\to\infty} B_n(t-x;x;\lambda) = 0; \tag{2.8}
$$

$$
\lim_{n\to\infty} nB_n((t-x)^2;x;\lambda) = x(1-x); \tag{2.9}
$$

$$\lim_{n\to\infty} n^2 B_n((t-x)^4; x; \lambda) = 3x^2(1-x)^2. \tag{2.10}$$

***Proof*** Suppose that $\lambda_i$ $(i = 1, \dots n)$ are the shape parameters with the property (1.3) and $x \in [0, 1]$. It follows from (2.5) that

$$\begin{aligned}\frac{(1-x)}{n}\mathcal{R}_0(x) &= \frac{1}{n}\sum_{i=1}^{[\frac{n}{2}]}(1-x)\frac{i^0\lambda_i}{\binom{n}{i}}b_{n,i}(x) - \frac{1}{n}\sum_{i=[\frac{n}{2}]+1}^{n}(1-x)\frac{i^0\lambda_i}{\binom{n}{i}}b_{n,i}(x)\\ &\le \frac{1}{n}\sum_{i=1}^{n+1}\frac{i}{n+1}b_{n+1,i}(x) = \frac{x}{n}.\end{aligned} \tag{2.11}$$

Conversely,

$$\begin{aligned}\frac{(1-x)\mathcal{R}_0(x)}{n} &\ge \frac{1}{n}\sum_{i=1}^{n}\frac{-\binom{n}{i}}{\binom{n+1}{i}}b_{n+1,i}(x) + \frac{1}{n}\sum_{i=1}^{n}\frac{-\binom{n}{i}}{\binom{n+1}{i}}b_{n+1,i}(x)\\ &\ge -\frac{1}{n}\sum_{i=1}^{n+1}\frac{n+1-i}{n+1}b_{n+1,i}(x) \ge -\frac{(1-x)}{n}.\end{aligned} \tag{2.12}$$

Therefore, $\lim_{n\to\infty} B_n(t-x; x; \lambda) = 0$. To prove (2.9), first we compute

$$\begin{aligned}\frac{(1-x)\mathcal{R}_1(x)}{n^2} &= \frac{1}{n^2}\sum_{i=1}^{[\frac{n}{2}]}(1-x)\frac{i\lambda_i}{\binom{n}{i}}b_{n,i}(x) - \frac{1}{n^2}\sum_{i=[\frac{n}{2}]+1}^{n}(1-x)\frac{i\lambda_i}{\binom{n}{i}}b_{n,i}(x)\\ &\le \frac{1}{n^2}\sum_{i=1}^{n} i\frac{\binom{n}{i-1}}{\binom{n+1}{i}}b_{n+1,i}(x) + \frac{1}{n^2}\sum_{i=1}^{n} i\frac{\binom{n}{i-1}}{\binom{n+1}{i}}b_{n+1,i}(x)\\ &\le \frac{n+1}{n^2}\sum_{i=1}^{n+1}\frac{i^2}{(n+1)^2}b_{n+1,i}(x) = \frac{n+1}{n^2}\left\{x^2 + \frac{x(1-x)}{n}\right\}.\end{aligned} \tag{2.13}$$

Analogous to (2.12), we get

$$\begin{aligned}\frac{(1-x)\mathcal{R}_1(x)}{n^2} &\ge \frac{1-x}{n^2}\sum_{i=1}^{[\frac{n}{2}]} i\frac{-\binom{n}{i}}{\binom{n}{i}}b_{n,i}(x) + \frac{1-x}{n^2}\sum_{i=[\frac{n}{2}]+1}^{n} i\frac{-\binom{n}{i}}{\binom{n}{i}}b_{n,i}(x)\\ &\ge -\frac{n(1-x)}{n^2}\sum_{i=1}^{n}\frac{i}{n}b_{n,i}(x) = -\frac{x(1-x)}{n}.\end{aligned} \tag{2.14}$$

Substituting the inequalities (2.11), (2.12), (2.13) and (2.14) into (2.7) gives us

$$nB_n((t-x)^2;x;\lambda) \le n\left\{\frac{x(1-x)}{n} + \frac{2x^2-x}{n^2} + \frac{2(n+1)x(1-x)}{n^3}\right\} \tag{2.15}$$

and

$$nB_n((t-x)^2;x;\lambda) \ge n\left\{\frac{x(1-x)}{n} + \frac{1-x}{n^2}\right\}. \tag{2.16}$$

Consequently, in view of (2.15) and (2.16), we have $\lim_{n\to\infty} nB_n((t-x)^2;x;\lambda) = x(1-x)$. The proof of (2.10) is essentially similar to the proof of (2.8) and (2.9); here we omit the detail. □

## 3 Approximation Properties of Operators $B_n(f;x;\lambda)$

Using well-known Bohman-Korovkin theorem (see, e.g., [22] or [23]), we get the uniform convergence of operators $B_n(f;x;\lambda)$.

**Theorem 3.1** *Let $f \in C[0,1]$, then we have*

$$\lim_{n\to\infty} \|B_n(f;\lambda) - f\|_{C[0,1]} = 0$$

*uniformly on* $[0,1]$, *where* $C[0,1]$ *denotes the space of all real-valued continuous functions on* $[0,1]$ *endowed with the norm* $\|f\|_{C[0,1]} = \sup_{x\in[0,1]} |f(x)|$.

***Proof*** Let us consider $e_j(t) = t^j$, where $t \in [0,1]$ and $j \in \{0,1,2\}$. According to Bohman-Korovkin theorem, it is sufficient to show that

$$\lim_{n\to\infty} \|B_n(e_j;x;\lambda) - e_j\|_{C[0,1]} = 0, \qquad j = 0,1,2.$$

By Lemma 2.1 we have

$$\lim_{n\to\infty} \|B_n(e_0;x;\lambda) - e_0\|_{C[0,1]} = 0 \quad \text{and} \quad \lim_{n\to\infty} \|B_n(e_1;x;\lambda) - e_1\|_{C[0,1]} = 0.$$

It is easy to show

$$\begin{aligned}\sup_{x\in[0,1]} \left|B_n(e_2;x;\lambda) - e_2(x)\right| &= \sup_{x\in[0,1]} \left\{\frac{x(1-x)}{n} + \frac{1-x}{n^2}[2\mathcal{R}_1(x) - \mathcal{R}_0(x)]\right\} \\ &\le \sup_{x\in[0,1]} \left\{\frac{x(1-x)}{n} + \frac{2(n+1)}{n^2}\left(x^2 + \frac{x(1-x)}{n}\right) - \frac{x}{n^2}\right\}\end{aligned}$$

and hence $\lim_{n\to\infty} \|B_n(e_2;x;\lambda) - e_2\|_{C[0,1]} = 0$. This implies $B_n(f;x;\lambda)$ converge uniformly to $f$ on $[0,1]$. □

Now, we recall some concepts and results concerning modulus of continuity and Peetre's K-functional (see [24]).

If $\delta > 0$, the modulus of continuity $w(f, \delta)$ of $f \in C[a, b]$ is defined by

$$w(f, \delta) := \sup\{|f(x) - f(y)| : x, y \in [a, b], |x - y| \leq \delta\}.$$

It is also well known that, for any $\delta > 0$ and each $x \in [a, b]$,

$$|f(x) - f(y)| \leq \omega(f, \delta)\left(\frac{|x - y|}{\delta} + 1\right). \tag{3.1}$$

The second order modulus of smoothness for $f \in C[0, 1]$ is defined as follows:

$$w_2(f, \sqrt{\delta}) := \sup_{0<h\leq\sqrt{\delta}} \sup_{x,\ x+2h\in[0,1]} \{|f(x + 2h) - 2f(x + h) + f(x)|\},$$

and the corresponding Peetre's $K$-functional [24] is defined by

$$K_2(f, \delta) = \inf\left\{||f - g||_{C[0,1]} + \delta||g''||_{C[0,1]} : g \in W^2[0, 1]\right\},$$

where $\delta > 0$ and $W^2[0, 1] = \{g \in C[0, 1] : g', g'' \in C[0, 1]\}$. It is also known that the inequality

$$K_2(f, \delta) \leq C\, w_2(f, \sqrt{\delta}) \tag{3.2}$$

holds for all $\delta > 0$, in which the absolute constant $C > 0$ is independent of $\delta$ and $f$ (see [25]).

First we establish a direct local approximation theorem for operators $B_n(f; x; \lambda)$ via second order modulus of smoothness and usual modulus of continuity.

**Theorem 3.2** *Let $f \in C[0, 1]$ and $x \in [0, 1]$. For the operators $B_n(f; x; \lambda)$, there exists an absolute constant $C$ such that*

$$|B_n(f; x; \lambda) - f(x)| \leq C\, w_2\Big(f, \frac{1}{2}\sigma_n(x)\Big) + w(f, \beta_n(x)),$$

*where $\beta_n(x) = B_n(t - x; x; \lambda)$, $\alpha_n(x) = B_n((t - x)^2; x; \lambda)$ and $\sigma_n(x) = \big(\alpha_n(x) + \beta_n^2(x)\big)^{\frac{1}{2}}$.*

***Proof*** For $x \in [0, 1]$, consider the operators

$$B_n^*(f; x; \lambda) = B_n(f; x; \lambda) + f(x) - f\Big(x + \frac{1 - x}{n}\mathcal{R}_0(x)\Big). \tag{3.3}$$

We observe that $B_n^*(1;x;\lambda)=1$ and $B_n^*(t;x;\lambda)=x$. That is to say that the operators $B_n^*(f;x;\lambda)$ preserve constants and linear functions, or equivalently $B_n^*(t-x;x;\lambda)=0$. Suppose $g\in W^2[0,1]$ and $t,x\in[0,1]$, by Taylor's expansion, we have

$$g(t)=g(x)+(t-x)g'(x)+\int_x^t (t-u)g''(u)du. \tag{3.4}$$

If we apply $B_n^*(f;x;\lambda)$ to both sides of (3.4), we obtain

$$\begin{aligned} B_n^*(g;x;\lambda)-g(x) &= g'(x)B_n^*(t-x;x;\lambda)+B_n^*\Big(\int_x^t (t-u)g''(u)du;x;\lambda\Big)\\ &= B_n\Big(\int_x^t (t-u)g''(u)du;x;\lambda\Big)-\int_x^{x+\beta_n}\big(x+\beta_n-u\big)g''(u)du, \end{aligned}$$

where $\beta_n=\frac{1-x}{n}\mathcal{R}_0(x)$. Therefore,

$$\begin{aligned} |B_n^*(g;x;\lambda)-g(x)| &\le B_n\Big(\Big|\int_x^t |t-u|\,|g''(u)|du\Big|;x;\lambda\Big)-\int_x^{x+\beta_n}\big|x+\beta_n-u\big|\,|g''(u)|\,du\\ &\le \|g''\|_{C[0,1]}\big(B_n((t-x)^2;x;\lambda)+\big(B_n(t-x;x;\lambda)\big)^2\big). \end{aligned}$$

In view of (3.3) we obtain

$$\|B_n^*(g;x;\lambda)\|_{C[0,1]}\le\|B_n(g;x;\lambda)\|_{C[0,1]}+\|g(x)\|_{C[0,1]}+\|g(x+\beta_n)\|_{C[0,1]}\le\|3g\|_{C[0,1]}. \tag{3.5}$$

Now, for $f\in C[0,1]$ and $g\in W^2[0,1]$, using (3.3) and (3.5) we get

$$\begin{aligned} |B_n(f;x;\lambda)-f(x)| &\le |B_n^*(f-g;x;\lambda)|+|B_n^*(g;x;\lambda)-g(x)|+|g(x)-f(x)|+|f(x+\beta_n)-f(x)|\\ &\le 4\|f-g\|_{C[0,1]}+\sigma_n^2(x)\|g''\|_{C[0,1]}+w(f,\beta_n). \end{aligned}$$

Hence, taking infimum on the right-hand side of the above inequality over all $g\in W^2[0,1]$ and using inequality (3.2), we get

$$|B_n(f;x;\lambda)-f(x)|\le 4K_2\Big(f,\frac{\sigma_n^2(x)}{4}\Big)+w(f,\beta_n)\le C\,w_2\Big(f,\frac{1}{2}\sigma_n(x)\Big)+\omega(f,\beta_n).$$

Thus the proof is complete. □

Now we obtain global approximation formula in terms of Ditzian-Totik uniform modulus of smoothness of first and second order (see [26]) defined by

$$\omega_\xi(f,\delta):=\sup_{0<|h|\le\delta}\ \sup_{x,\,x+h\xi(x)\in[0,1]}\{|f(x+h\xi(x))-f(x)|\}$$

and

$$\omega_2^{\phi}(f,\delta) := \sup_{0<|h|\le\delta}\ \sup_{x,\ x\pm h\phi(x)\in[0,1]} \{|f(x+h\phi(x)) - 2f(x) + f(x-h\phi(x))|\},$$

respectively, where $\phi$ is an admissible step-weight function on $[a,b]$, i.e., $\phi(x) = [(x-a)(b-x)]^{1/2}$ if $x \in [a,b]$. Corresponding $K$-functional is

$$K_{2,\phi(x)}(f,\delta) = \inf_{g\in W^2(\phi)} \left\{||f-g||_{C[0,1]} + \delta||\phi^2 g''||_{C[0,1]} : g \in C^2[0,1]\right\},$$

where $\delta > 0$, $W^2(\phi) = \{g \in C[0,1] : g' \in AC[0,1],\ \ \phi^2 g'' \in C[0,1]\}$ and $C^2[0,1] = \{g \in C[0,1] : g', g'' \in C[0,1]\}$. Here, $g' \in AC[0,1]$ means that $g'$ is absolutely continuous on $[0,1]$. It is known (see [26]) that there exists an absolute constant $C > 0$, such that

$$C^{-1}\omega_2^{\phi}(f,\sqrt{\delta}) \le K_{2,\phi(x)}(f,\delta) \le C\omega_2^{\phi}(f,\sqrt{\delta}). \tag{3.6}$$

**Theorem 3.3** *Let $\phi$ ($\phi \ne 0$) be an admissible step-weight function of Ditzian-Totik modulus of smoothness such that $\phi^2$ is concave and $f \in C[0,1]$. Then, for any $x \in [0,1]$ and $C > 0$, we have*

$$|B_n(f;x;\lambda) - f(x)| \le C\omega_2^{\phi}\left(f, \frac{\sigma_n(x)}{2\phi(x)}\right) + \omega_{\xi}\left(f, \frac{\beta_n(x)}{\xi(x)}\right),$$

*where $\beta_n$ and $\sigma_n$ are defined as in Theorem 3.2.*

***Proof*** We consider the operator $B_n^*(f;x;\lambda)$ defined in (3.3). Let $u = \rho x + (1-\rho)t$, $\rho \in [0,1]$. Since $\phi^2$ is concave on $[0,1]$, it follows that $\phi^2(u) \ge \rho\phi^2(x) + (1-\rho)\phi^2(t)$ and hence

$$\frac{|t-u|}{\phi^2(u)} \le \frac{\rho|x-t|}{\rho\phi^2(x) + (1-\rho)\phi^2(t)} \le \frac{|t-x|}{\phi^2(x)}. \tag{3.7}$$

So

$$\begin{aligned} |B_n^*(f;x;\lambda) - f(x)| &\le |B_n^*(f-g;x;\lambda)| + |B_n^*(g;x;\lambda) - g(x)| + |f(x) - g(x)| \\ &\le 4\|f-g\|_{C[0,1]} + |B_n^*(g;x;\lambda) - g(x)|. \end{aligned} \tag{3.8}$$

Applying Taylor's formula we obtain

$$\begin{aligned}|B_n^*(g;x;\lambda)-g(x)| &\le B_n\left(\left|\int_x^t |t-u|\,|g''(u)|du\right|;x;\lambda\right)+\left|\int_x^{x+\beta_n}|x+\beta_n-u|\,|g''(u)|\,du\right| \qquad (3.9)\\ &\le \|\phi^2g''\|_{C[0,1]}B_n\left(\left|\int_x^t \frac{|t-u|}{\phi^2(u)}du\right|;x;\lambda\right)+\|\phi^2g''\|_{C[0,1]}\left|\int_x^{x+\beta_n}\frac{|x+\beta_n-u|}{\phi^2(u)}\,du\right|\\ &\le \phi^{-2}(x)\|\phi^2g''\|_{C[0,1]}\alpha_n(x)+\phi^{-2}(x)\|\phi^2g''\|_{C[0,1]}\beta_n^2(x).\end{aligned}$$

From (3.8), (3.9) and by using definition of $K$-functional along with the relation (3.6),

$$\begin{aligned}|B_n^*(f;x;\lambda)-f(x)| &\le 4\|f-g\|_{C[0,1]}+\phi^{-2}(x)\|\phi^2g''\|_{C[0,1]}\big(\alpha_n(x)+\beta_n^2(x)\big)\\ &\le C\omega_2^{\phi}\left(f,\frac{\sqrt{\alpha_n(x)+\beta_n^2(x)}}{2\phi(x)}\right).\end{aligned}$$

On the other hand, from the Ditzian-Totik uniform modulus of smoothness of first order we have $|f(x+\beta_n)-f(x)|=\left|f\left(x+\xi(x)\frac{\beta_n}{\xi(x)}\right)-f(x)\right|\le\omega_\xi\left(f,\frac{\beta_n}{\xi(x)}\right)$. Hence

$$|B_n(f;x;\lambda)-f(x)|\le|B_n^*(f;x;\lambda)-f(x)|+\left|f(x+\beta_n)-f(x)\right|\le C\omega_2^{\phi}\left(f,\frac{\sigma_n(x)}{2\phi(x)}\right)+\omega_\xi\left(f,\frac{\beta_n(x)}{\xi(x)}\right).$$

This completes the proof. □

Now we give a local direct estimate of the rate of convergence with the help of Lipschitz-type function involving two parameters (see [27]).
For $k_1\ge 0$, $k_2>0$, we define

$$Lip_M^{(k_1,k_2)}(\eta):=\left\{f\in C[0,1]:|f(t)-f(x)|\le M\frac{|t-x|^{\eta}}{(k_1x^2+k_2x+t)^{\frac{\eta}{2}}};\ x\in(0,1],t\in[0,1]\right\},$$

where $M$ is a positive constant and $\eta\in(0,1]$.

**Theorem 3.4** *Let $f\in Lip_M^{(k_1,k_2)}(\eta)$, $\eta\in(0,1]$. Then, for all $x\in(0,1]$ we have*

$$|B_n(f;x;\lambda)-f(x)|\le M\left(\frac{\alpha_n(x)}{k_1x^2+k_2x}\right)^{\frac{\eta}{2}},$$

*where $\alpha_n(x)$ is defined as in Theorem 3.2.*

***Proof*** Let $f\in Lip_M^{(k_1,k_2)}(\eta)$ and $\eta\in(0,1]$. It is easy to see that the statement is true for $\eta=1$. In fact, for $f\in Lip_M^{(k_1,k_2)}(1)$, we have

$$\begin{aligned}|B_n(f;x;\lambda)-f(x)| &\leq |B_n(|f(t)-f(x)|;x;\lambda)|+f(x)\,|B_n(1;x;\lambda)-1|\\ &\leq \sum_{i=0}^{n}\left|f\left(\frac{i}{n}\right)-f(x)\right| a_{n,i}(\lambda;x)\\ &\leq M\sum_{i=0}^{n}\frac{|\frac{i}{n}-x|}{(k_1x^2+k_2x+t)^{\frac{1}{2}}}\,a_{n,i}(\lambda;x).\end{aligned}$$

Using the fact that $\frac{1}{\sqrt{k_1x^2+k_2x+t}} \leq \frac{1}{\sqrt{k_1x^2+k_2x}}$ for $k_1 \geq 0, k_2 > 0$ and applying Cauchy-Schwarz inequality, we get

$$\begin{aligned}|B_n(f;x;\lambda)-f(x)| &\leq \frac{M}{\sqrt{k_1x^2+k_2x}}\sum_{i=0}^{n}\left|\frac{i}{n}-x\right| a_{n,i}(\lambda;x)\\ &= \frac{M}{\sqrt{k_1x^2+k_2x}}|B_n(t-x;x;\lambda)| \leq M\left(\frac{|\alpha_n(x)|}{k_1x^2+k_2x}\right)^{\frac{1}{2}}.\end{aligned}$$

Therefore the statement is true for $\eta = 1$. Now we show that the statement is true for $\eta \in (0, 1]$. Using the monotonocity of the operators $B_n(f;x;\lambda)$ and applying Hölder's inequality two times with $u = \frac{2}{\eta}$, $v = \frac{2}{2-\eta}$, we deduce

$$\begin{aligned}|B_n(f;x;\lambda)-f(x)| &\leq \sum_{i=0}^{n}\left|f\left(\frac{i}{n}\right)-f(x)\right| a_{n,i}(\lambda;x)\\ &\leq \left(\sum_{i=0}^{n}\left|f\left(\frac{i}{n}\right)-f(x)\right|^{\frac{2}{\eta}} a_{n,i}(\lambda;x)\right)^{\frac{\eta}{2}}\left(\sum_{i=0}^{n}a_{n,i}(\lambda;x)\right)^{\frac{2-\eta}{2}}\\ &\leq M\left(\sum_{i=0}^{n}\frac{\left(\frac{i}{n}-x\right)^2 a_{n,i}(\lambda;x)}{\frac{i}{n}+k_1x^2+k_2x}\right)^{\frac{\eta}{2}}\\ &\leq \frac{M}{(k_1x^2+k_2x)^{\frac{\eta}{2}}}\left\{\sum_{i=0}^{n}\left(\frac{i}{n}-x\right)^2 a_{n,i}(\lambda;x)\right\}^{\frac{\eta}{2}}\\ &\leq \frac{M}{(k_1x^2+k_2x+t)^{\frac{\eta}{2}}}(B_n((t-x)^2;x;\lambda))^{\frac{\eta}{2}} = M\left(\frac{\alpha_n(x)}{k_1x^2+k_2x}\right)^{\frac{\eta}{2}}.\end{aligned}$$

Thus the proof is completed. □

**Theorem 3.5** *Let* $f \in C^1[0,1]$. *For any* $x \in [0,1]$, *the following inequality holds:*

$$|B_n(f;x;\lambda)-f(x)| \leq |\beta_n(x)|\,|f'(x)| + 2\sqrt{\alpha_n(x)}w\left(f', \sqrt{\alpha_n(x)}\,\right).$$

***Proof*** For any $t \in [0,1]$, $x \in [0,1]$, we have

$$f(t)-f(x) = (t-x)f'(x) + \int_x^t (f'(u)-f'(x))du.$$

Applying $B_n(f; x; \lambda)$ on both sides of the above relation, we obtain

$$B_n(f(t) - f(x); x; \lambda) = f'(x)B_n(t - x; x; \lambda) + B_n\left(\int_x^t (f'(u) - f'(x))du; x; \lambda\right).$$

It is well known that for any $\delta > 0$ and each $u \in [0, 1]$,

$$|f(u) - f(x)| \leq w(f, \delta)\left(\frac{|u - x|}{\delta} + 1\right), \quad f \in C[0, 1].$$

With the above inequality we have

$$\left|\int_x^t (f'(u) - f'(x))du\right| \leq w(f', \delta)\left(\frac{(t - x)^2}{\delta} + |t - x|\right).$$

Thus,

$$|B_n(f; x; \lambda) - f(x)| \leq |f'(x)|\,|B_n(t - x; x; \lambda)| + w(f', \delta)\left\{\frac{1}{\delta}B_n((t - x)^2; x; \lambda) + B_n(t - x; x; \lambda)\right\}. \tag{3.10}$$

Applying Cauchy-Schwarz inequality on the right-hand side of (3.10), we have

$$|B_n(f; x; \lambda) - f(x)| \leq f'(x)|\beta_n(x)| + w(f', \delta)\left\{\frac{1}{\delta}\sqrt{B_n((t - x)^2; x; \lambda)} + 1\right\}\sqrt{B_n(|t - x|; x; \lambda)}.$$

Choosing $\delta = \sqrt{\alpha_n(x)}$, we get the desired result. □

## 4 Statistical Approximation Properties by Weighted Mean Matrix Method

In this section, we present statistical approximation properties by weighted mean matrix method for operators $B_n(t - x; x; \lambda)$ in terms of Korovkin-type statistical approximation theorem proved in [28] (see also [29–32]).

Let $K \subseteq \mathbb{N}_0 := \mathbb{N} \cup \{0\}$ and $K_n = \{k \leq n : k \in K\}$. Then $\delta(K) = \lim_n \frac{1}{n}|K_n|$ is called *natural density* of $K_n$ provided that the limit exists, where the vertical bars denote cardinality of the enclosed set (see [33]). A number sequence $x = (x_n)$ is called statistically convergent to a number $L$, denoted by $st\text{-}\lim_n x = L$, if, for each $\epsilon > 0$,

$$\delta\{n : n \in \mathbb{N} \text{ and } |x_n - L| \geqq \epsilon\} = 0.$$

For a given non-negative infinite summability matrix $A = (a_{nk}), n, k \in \mathbb{N}$, the $A$-transform of $x$ denoted by $Ax := \{(Ax)_n\}$ is defined as $(Ax)_n = \sum_{k=0}^{\infty} a_{nk}x_k$ provided the series converges for every $n \in \mathbb{N}_0$. We say that $A$ is regular method, if $\lim_n (Ax)_n = L$ whenever $\lim_n x_n = L$. Then $x = (x_n)$ is said to be $A$-statistically convergent to $L$, denoted by $\mathrm{st}_A$-$\lim x = L$, provided that for each $\epsilon > 0$, $\lim_{n\to\infty} \sum_{k:|x_k-L|\geqq\epsilon} a_{nk} = 0$. If we take $A = (C_1)$, the Cesaro matrix of order one, then $A$-statistical convergence is reduced to ordinary statistical convergence which was introduced by Gadjiev and Orhan [28]. Also, taking $A = I$, the identity matrix, we get classical convergence. It is easy to check that every convergent sequence is statistically convergent to the same limit but not conversely. Read more on the subject of $A$-statistical convergence in [34–37] and references cited therein.

Recently Mohiuddine introduced an extended form of $A$-statistical convergence in [38], namely weighted $A$-statistical convergence by using a non-negative weighted regular matrix. He also gave a new characterization in terms of the weighted regular matrix and proved Korovkin type approximation theorem through statistically weighted $A$-summable sequences of real or complex numbers.

Let us assume that $p = (p_n)$ is a sequence of non-negative numbers such that $p_0 > 0$ and $P_n = \sum_{k=0}^{n} p_k \to \infty$, as $n \to \infty$. Then $x = (x_n)$ is said to be weighted $A$-statistically convergent to $L$, if, for every $\varepsilon > 0$,

$$\lim_{n\to\infty} \frac{1}{P_n} \sum_{k=0}^{n} p_k \sum_{l:|x_l-L|\geqq\epsilon} a_{kl} = 0.$$

In this case we denote it by $S_A^{\widetilde{N}} - \lim x = L$.

Obviously, the concept of weighted $A$-statistical convergence generalizes the $A$-statistical convergence, which we recover by putting $p_n = 1$ for all $n \in \mathbb{N}$.

Now, we prove a Korovkin-type statistical approximation theorem for the operators $B_n(f; x; \lambda)$ (2.1).

**Theorem 4.1** *Let $A = (a_{nk})$, $n, k \in \mathbb{N}$ be a weighted non-negative regular summability matrix. Let $p = (p_n)$ be a sequence of non-negative numbers such that $p_0 > 0$ and $P_n = \sum_{k=0}^{n} p_k \to \infty$, as $n \to \infty$. For any $f \in C[0, 1]$, we have*

$$S_A^{\widetilde{N}} - \lim_{n\to\infty} \|B_n(f; x; \lambda) - f(x)\|_{C[0,1]} = 0. \tag{4.1}$$

***Proof*** Let us consider $e_j(x) = x^j$, where $x \in [0, 1]$ and $j \in \{0, 1, 2\}$. It is sufficient to verify that

$$S_A^{\widetilde{N}} - \lim_{n\to\infty} \|B_n(e_j; x; \lambda) - e_j\|_{C[0,1]} = 0, \qquad j = 0, 1, 2.$$

From Lemma 2.1, it is clear that

$$S_A^{\widetilde{N}} - \lim_{n\to\infty} \|B_n(e_0; x; \lambda) - e_0\|_{C[0,1]} = 0. \tag{4.2}$$

In view of Lemma 2.1 and Corollary 2.1, we obtain

$$\|B_n(e_1;x;\lambda)-e_1\|_{C[0,1]} = \sup_{x\in[0,1]}\left|x+\frac{1-x}{n}\mathcal{R}_0(x)-x\right| \leq \frac{1}{n}.$$

For a given $\epsilon' > 0$, we choose a number $\epsilon > 0$ such that $\epsilon < \epsilon'$. Let us define the following sets:

$$\mathcal{E} := \left\{n \in \mathbb{N} : \|B_n(e_1;x;\lambda)-e_1\|_{C[0,1]} \geqq \epsilon'\right\},$$
$$\mathcal{E}_1 := \left\{n \in \mathbb{N} : \frac{1}{n} \geqq \epsilon - \epsilon'\right\}.$$

Then, it is obvious that $\mathcal{E} \subset \mathcal{E}_1$ which guarantees, for all $n \in \mathbb{N}$, that

$$\frac{1}{P_n}\sum_{k=0}^{n} p_k \sum_{l\in\mathcal{E}} a_{kl} \leq \frac{1}{P_n}\sum_{k=0}^{n} p_k \sum_{l\in\mathcal{E}_1} a_{kl}. \tag{4.3}$$

By letting $n \to \infty$ in (4.3) one can derive that

$$S_A^{\widetilde{N}} - \lim_{n\to\infty} \|B_n(e_1;x;\lambda)-e_1\|_{C[0,1]} = 0. \tag{4.4}$$

Again, from Lemma 2.1 and Corollary 2.1, we have

$$\|B_n(e_2;x;\lambda)-e_2\|_{C[0,1]} = \sup_{x\in[0,1]}\left|\frac{x(1-x)}{n}+\frac{1-x}{n^2}\left[2\mathcal{R}_1(x)-\mathcal{R}_0(x)\right]\right|$$

and hence

$$S_A^{\widetilde{N}} - \lim_{n\to\infty} \|B_n(e_2;x;\lambda)-e_2\|_{C[0,1]} = 0. \tag{4.5}$$

Finally combining (4.2), (4.4) and (4.5) we obtain the desired result. □

## *4.1 Rate of Weighted A-statistical convergence*

Here we estimate the rate of weighted $A$-statistical convergence of the operators $B_n(f;x;\lambda)$. We first present the following definition.

**Definition 4.1** Let $A = (a_{nk})$ be a weighted non-negative regular summability matrix and let $p = (p_n)$ be a sequence of non-negative numbers such that $p_0 > 0$ and $P_n = \sum_{k=0}^{n} p_k \to \infty$ as $n \to \infty$. Also let $(u_n)$ be a positive non-decreasing sequence. We say that a sequence $x = (x_n)$ is weighted $A$-statistically convergent to $L$ with the rate $o(u_n)$ if

$$\lim_{n\to\infty}\frac{1}{u_nP_n}\sum_{k=0}^{n}p_k\sum_{l:|x_l-L|\geqq\epsilon}a_{kl}=0.$$

In this case, we can write

$$[stat_A,p_n]-o\,(u_n)=x_n-L.$$

**Theorem 4.2** *Let $A=(a_{nk})$ be a weighted non-negative regular summability matrix. Suppose that the following condition holds:*

$w(f,\alpha_n')=[stat_A,p_n]-o\,(u_n)$ *on* $[0,1]$, *where* $\alpha_n'=\|B_n((t-x)^2;x;\lambda)\|_{C[0,1]}^{1/2}$.

*Then, for every bounded $f\in C[0,1]$, we have*

$$\|B_n(f;x;\lambda)-f(x)\|_{C[0,1]}=[stat_A,p_n]-o\,(u_n).$$

***Proof*** Let $f\in C_{[0,1]}$. For any $x,t\in[0,1]$, we have

$$\begin{aligned}|B_n(f;x;\lambda)-f(x)|&\le|B_n\big(|f(t)-f(x)|;x;\lambda\big)+N\,|B_n(1;x;\lambda)-1|\\&\le\omega(f,\delta)B_n\left(\frac{|t-x|}{\delta}+1;x;\lambda\right)\\&=\omega(f,\delta)B_n(1;x;\lambda)+\omega(f,\delta)\frac{1}{\delta^2}B_n\left((t-x)^2;x;\lambda\right)\end{aligned}$$

where $N=\sup_{x\in[0,1]}|f(x)|$. Let $\delta:=\alpha_n'$ for all $n\in\mathbb{N}$. Taking the supremum over $x\in[0,\infty)$ on both sides, we get

$$\|B_n(f;x;\lambda)-f(x)\|_{C[0,1]}\le\omega(f,\alpha_n')+\omega(f,\alpha_n')\frac{1}{(\alpha_n')^2}\|B_n((t-x)^2;x;\lambda)\|_{C[0,1]}=2\omega(f,\alpha_n').$$

For a given $\epsilon>0$, we consider the following sets:

$$\mathcal{U}=\left\{n:\|B_n(f;x;\lambda)-f(x)\|_{C[0,1]}\ge\epsilon\right\},\quad\mathcal{V}=\left\{n:\omega(f,\alpha_n')\ge\frac{\epsilon}{2}\right\}.$$

It is clear that

$$\frac{1}{u_nP_n}\sum_{k=0}^{n}\sum_{l\in\mathcal{U}}p_ka_{kl}\le\frac{1}{u_nP_n}\sum_{k=0}^{n}\sum_{l\in\mathcal{V}}p_ka_{kl}.$$

Hence by the hypothesis we are led to the fact that

$$\|B_n(f;x;\lambda)-f(x)\|_{C[0,1]}=[stat_A,p_n]-o\,(u_n),$$

as asserted by Theorem 4.2. □

## 5 Voronovskaja Type Theorems

In this section, we establish Voronovskaja type results with the help of $B_n(f; x; \lambda)$ family of linear operators defined by (2.1).

**Theorem 5.1** *For every $f \in C[0, 1]$ such that $f', f'' \in C[0, 1]$, and for each $x \in [0, 1]$, we have*

$$\lim_{n\to\infty} n\{B_n(f; x; \lambda) - f(x)\} = (1 - x)\, \mathcal{R}_0(x) f'(x) + x(1 - x)\frac{f''(x)}{2}$$

*uniformly on $[0, 1]$, where $\mathcal{R}_0(x)$ is defined as in Lemma 2.1.*

***Proof*** Let $x \in [0, 1]$. By Taylor's expansion theorem of function $f$ in $C[0, 1]$ we obtain

$$f(t) = f(x) + (t - x) f'(x) + \frac{1}{2}(t - x)^2 f''(x) + (t - x)^2\, r_x(t), \tag{5.1}$$

where $r_x(t)$ is Peano form of the remainder, $r_x \in C[0, 1]$ and $r_x(t) \to 0$ as $t \to x$. Operating $B_n(f; x; \lambda)$ to the identity (5.1), we get

$$B_n(f; x; \lambda) - f(x) = f'(x) B_n(t - x; x; \lambda) + \frac{f''(x)}{2} B_n((t - x)^2; x; \lambda) + B_n((t - x)^2 r_x(t); x; \lambda).$$

Using Cauchy-Schwarz inequality, we have

$$n B_n((t - x)^2 r_x(t); x; \lambda) \le \sqrt{n^2 B_n((t - x)^4; x; \lambda)}\sqrt{B_n(r_x^2(t); x; \lambda)}. \tag{5.2}$$

On the other hand, using (2.10), we have

$$\lim_{n\to\infty} \sqrt{n^2 B_n((t - x)^4; x; \lambda)} = 3x^2(1 - x)^2.$$

By Theorem 3.1, it follows that $\lim_n B_n(r_x^2(t); x; \lambda) = 0$ and hence $\lim_n n\{B_n((t - x)^2 r_x(t); x; \lambda)\} = 0$. Therefore

$$\lim_{n\to\infty} n\{B_n(f; x; \lambda) - f(x)\} = \lim_{n\to\infty} n\big\{B_n(t - x; x; \lambda) f'(x) + \frac{f''(x)}{2} B_n((t - x)^2; x; \lambda) + B_n((t - x)^2 r_x(t); x; \lambda)\big\}.$$

The result follows immediately by applying Theorem 2.1. □

Now, based on the notion of weighted $A$-statistical convergence introduced in the preceding section, we prove a Voronovskaja type approximation theorem by the help of $\widetilde{B}_n(f; x; \lambda)$ family of linear operators.

**Theorem 5.2** *Let $A = (a_{nk})$ be a weighted non-negative regular summability matrix and let $(x_n)$ be a sequence of real numbers such that $S_A^N - \lim x_n = 0$. Also let $\widetilde{B}_n(f; x; \lambda)$ be a sequence of positive linear operators acting from $C[0, 1]$ into itself defined by*

$$\widetilde{B}_n(f; x; \lambda) = (1 + x_n)B_n(f; x; \lambda),$$

*where $B_n(f; x; \lambda)$ is defined as in (2.1). Then, for every $f \in C[0, 1]$ such that $f', f'' \in C[0, 1]$,*

$$S_A^{\widetilde{N}} - \lim_{n\to\infty} n\{\widetilde{B}_n(f; x; \lambda) - f(x)\} = f'(x)(1 - x)\mathcal{R}_0(x) + \frac{f''(x)}{2}x(1 - x).$$

***Proof*** Let $x \in [0, 1]$ and $f', f'' \in C[0, 1]$. By using the same argument as in Theorem 5.1 and applying $\widetilde{B}_n(f; x; \lambda)$ to both sides of (5.1), we have

$$\widetilde{B}_n(f; x; \lambda) - f(x) = f'(x)\widetilde{B}_n(t - x; x; \lambda) + \frac{f''(x)}{2}\widetilde{B}_n((t - x)^2; x; \lambda) + \widetilde{B}_n((t - x)^2 r_x(t); x; \lambda), \quad (5.3)$$

which yields

$$\begin{aligned} n\{\widetilde{B}_n(f; x; \lambda) - f(x)\} &= nf'(x)(1 + x_n)B_n(t - x; x; \lambda) + \frac{n}{2}f''(x)(1 + x_n)B_n((t - x)^2; x; \lambda) \\ &\quad + n(1 + x_n)B_n((t - x)^2 r_x(t); x; \lambda). \end{aligned} \quad (5.4)$$

Using (2.11) and (2.15), we deduce that

$$nB_n(t - x; x; \lambda) \le (1 - x)\mathcal{R}_0(x), \quad nB_n((t - x)^2; x; \lambda) \le x(1 - x)(1 - 2\mathcal{R}_0(x)) + \frac{1 - x}{n}[2\mathcal{R}_1(x) - \mathcal{R}_0(x)],$$

and we obtain

$$\begin{aligned} &\left| n\{\widetilde{B}_n(f; x; \lambda) - f(x)\} - \left\{ nf'(x)B_n(t - x; x; \lambda) + n\frac{f''(x)}{2}B_n((t - x)^2; x; \lambda) \right\} \right| \\ &\le f'(x)x_n(1 - x)\mathcal{R}_0(x) + \frac{f''(x)}{2}x_n\left[x(1 - x)(1 - 2\mathcal{R}_0(x)) + \frac{1 - x}{n}(2\mathcal{R}_1(x) - \mathcal{R}_0(x))\right] \\ &\quad + n(1 + x_n)B_n((t - x)^2 r_x(t); x; \lambda) \\ &\le N_0 x_n(1 - x)\mathcal{R}_0 + \frac{1}{2}N_1 x_n\left[x(1 - x)(1 - 2\mathcal{R}_0) + \frac{1 - x}{n}(2\mathcal{R}_1 - \mathcal{R}_0)\right] + n(1 + x_n)B_n((t - x)^2 r_x; x; \lambda) \\ &\le N_0 x_n \mathcal{R}_0(x) + x_n\frac{N_1}{2}\left[(1 - 2\mathcal{R}_0(x)) + \frac{1}{n}(2\mathcal{R}_1(x) - \mathcal{R}_0(x))\right] + n(1 + x_n)B_n((t - x)^2 r_x(t); x; \lambda), \end{aligned}$$

where $N_0 = \sup_{x\in[0,1]} |f'(x)|$ and $N_1 = \sup_{x\in[0,1]} |f''(x)|$. Further, in view of Theorem 4.1, it follows that

$$S_A^{\widetilde{N}} - \lim_{n\to\infty} n(1 + x_n)B_n((t - x)^2 r_x(t); x; \lambda) = 0.$$

Using the hypothesis $S_A^{\widetilde{N}} - \lim x_n = 0$ and taking into account Theorem 2.1, we get the desired result. □

In the next theorem, we obtain a quantitative Voronovskaja type theorem for $B_n(f; x; \lambda)$ using the Ditzian-Totik modulus of smoothness of first order defined as follows (see [26, 39]):

$$\omega_\phi(f, \delta) := \sup_{0<|h|\leq\delta} \left\{ \left| f\left(x + \frac{h\phi(x)}{2}\right) - f\left(x - \frac{h\phi(x)}{2}\right) \right|, x \pm \frac{h\phi(x)}{2} \in [0, 1] \right\},$$

where $\phi(x) = \sqrt{x(1-x)}$ and $f \in C[0, 1]$, and the corresponding Peetre's $K$-functional is defined by

$$K_\phi(f, \delta) = \inf_{g\in W_\phi[0,1]} \left\{ ||f - g|| + \delta||\phi g'|| : g \in C^1[0, 1], \delta > 0 \right\},$$

where $W_\phi[0, 1] = \{g : g \in AC_{loc}[0, 1], \ \|\phi g'\| < \infty\}$ and $AC_{loc}[0, 1]$ is the class of absolutely continuous functions defined on $[a, b] \subset [0, 1]$. Besides, from [26], there exists a constant $C > 0$ such that

$$K_\phi(f, \delta) \leq C\ \omega_\phi(f, \delta).$$

**Theorem 5.3** *Suppose that* $f \in C[0, 1]$ *such that* $f', f'' \in C[0, 1]$. *Then, for every* $x \in [0, 1]$ *and sufficiently large n, we have*

$$\left| B_n(f; x; \lambda) - f(x) - \beta_n f'(x) - \left(\frac{\alpha_n + 1}{2}\right) f''(x) \right| \leq \frac{C}{n} \phi^2(x) \omega_\phi\left(f'', \frac{1}{n^2}\right),$$

*where C is a positive constant,* $\alpha_n$ *and* $\beta_n$ *are as indicated in Theorem 3.2.*

***Proof*** First, it is easy to see that

$$f(t) - f(x) - (t - x)f'(x) = \int_x^t (t - u)f''(u)du$$

for $f \in C[0, 1]$. This yields

$$f(t) - f(x) - (t - x)f'(x) - \frac{f''(x)}{2}\left((t - x)^2 + 1\right) \leq \int_x^t (t - u)[f''(u) - f''(x)]du. \tag{5.5}$$

If we apply $B_n(f; x; \lambda)$ to both sides of (5.5), we have

$$\left| B_n(f; x; \lambda) - f(x) - B_n((t - x); x; \lambda)f'(x) - \frac{f''(x)}{2}\left(B_n((t - x)^2; x; \lambda) + B_n(1; x; \lambda)\right) \right|$$
$$\leq B_n\left( \left| \int_x^t |t - u|\ |f''(u) - f''(x)|\ du \right|; x; \lambda \right). \tag{5.6}$$

Take into consideration that the quantity on the right-hand side of (5.6) can be estimated as follows (see [40]):

$$\left|\int_x^t |t-u|\,|f''(u)-f''(x)|\,du\right| \leq 2\|f''-g\|(t-x)^2 + 2\|\phi g'\|\phi^{-1}(x)|t-x|^3, \tag{5.7}$$

where $g \in W_\phi[0,1]$. From Corollary 2.1, there exists $C > 0$ such that

$$B_n((t-x)^2;x;\lambda) \leq \frac{C}{2n}x(1-x) \quad \text{and} \quad B_n((t-x)^4;x;\lambda) \leq \frac{C}{2n^2}x^2(1-x)^2 \tag{5.8}$$

for sufficiently large $n$. From (5.6)–(5.8) and using Cauchy-Schwarz inequality, we conclude that

$$\begin{aligned}
&\left|B_n(f;x;\lambda) - f(x) - B_n((t-x);x;\lambda)f'(x) - \frac{f''(x)}{2}\left(B_n((t-x)^2;x;\lambda) + B_n(1;x;\lambda)\right)\right| \\
&\leq 2\|f''-g\|B_n((t-x)^2;x;\lambda) + 2\|\phi g'\|\phi^{-1}(x)B_n(|t-x|^3;x;\lambda) \\
&\leq \frac{C}{n}x(1-x)\|f''-g\| + 2\|\phi g'\|\phi^{-1}(x)\{B_n((t-x)^2;x;\lambda)\}^{1/2}\{B_n((t-x)^4;x;\lambda)\}^{1/2} \\
&\leq \frac{C}{n}\phi^2(x)\left\{\|f''-g\| + \frac{1}{n^2}\|\phi g'\|\right\}
\end{aligned}$$

and taking the infimum on the right-hand side over all $g \in W_\phi[0,1]$, we deduce

$$\left|B_n(f;x;\lambda) - f(x) - \beta_n f'(x) - \frac{\alpha_n+1}{2}f''(x)\right| \leq \frac{C}{n}\phi^2(x)\omega_\phi\left(f'', \frac{1}{n^2}\right).$$

This completes the proof of theorem. □

**Corollary 5.1** *If $f \in C[0,1]$ such that $f', f'' \in C[0,1]$, then*

$$\lim_{n\to\infty} n\left\{B_n(f;x;\lambda) - f(x) - \beta_n f'(x) - \left(\frac{\alpha_n+1}{2}\right)f''(x)\right\} = 0.$$

## 6 Graphical Results

In this section, we first focus on the behavior of generalized basis polynomials $a_{n,i}(\lambda;x)$ and give several illustrative examples with the help of Maple to verify the convergence behavior, computational efficiency and consistency of the generalized Bernstein operators.

In order to see the effect of shape parameters $\lambda$ to behavior of polynomials $a_{n,i}(\lambda;x)$, we choose certain parameters $\lambda$ below. In Fig. 1 (A), we present behavior of classical Bernstein basis polynomials $b_{n,i}(x)$ to see the difference. In Fig. 1 (B)–(F), we choose the following shape parameters:

$$(B)\ \lambda_i = \binom{n}{i-1} - \binom{n}{i},\ i = 1, 2, \ldots, [\tfrac{n}{2}],\quad \lambda_i = \binom{n}{i} - \binom{n}{i-1},\ i = [\tfrac{n}{2}] + 1, \ldots, n$$

$$(C)\ \lambda_i = \binom{n}{i-1},\ i = 1, 2, \ldots, [\tfrac{n}{2}],\quad \lambda_i = \binom{n}{i},\ i = [\tfrac{n}{2}] + 1, \ldots, n$$

$$(D)\ \lambda_i = -\binom{n}{i},\ i = 1, 2, \ldots, [\tfrac{n}{2}],\quad \lambda_i = \binom{n}{i},\ i = [\tfrac{n}{2}] + 1, \ldots, n$$

$$(E)\ \lambda_i = \binom{n}{i-1},\ i = 1, 2, \ldots, [\tfrac{n}{2}],\quad \lambda_i = -\binom{n}{i-1},\ i = [\tfrac{n}{2}] + 1, \ldots, n$$

$$(F)\ \lambda_i = -\binom{n}{i},\ i = 1, 2, \ldots, [\tfrac{n}{2}],\quad \lambda_i = -\binom{n}{i-1},\ i = [\tfrac{n}{2}] + 1, \ldots, n.$$

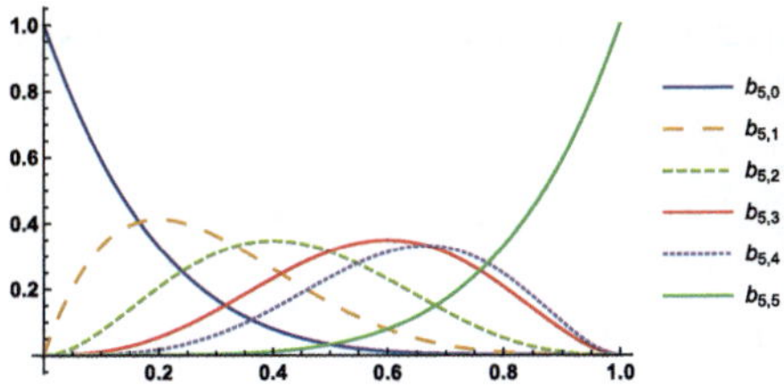


(A) $b_{5,i}(x)$ for $i = 0, 1, \cdots 5$

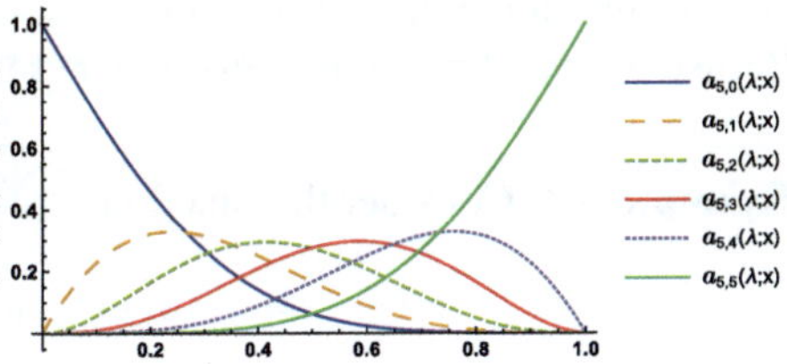


(B) $a_{5,i}(\lambda; x)$ with $\lambda_i = \binom{5}{i-1} - \binom{5}{i}$, $i = 1, 2, 3$ and $\lambda_i = \binom{5}{i} - \binom{5}{i-1}$, $i = 4, 5$

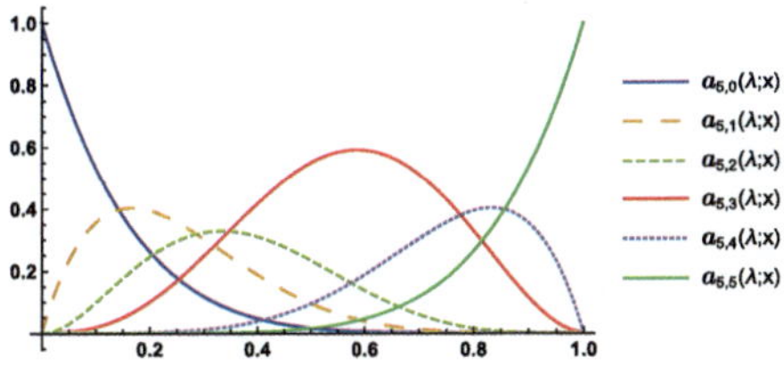


(C) $a_{5,i}(\lambda; x)$ with $\lambda_i = \binom{5}{i-1}$, $i = 1, 2, 3$ and $\lambda_i = \binom{5}{i}$, $i = 4, 5$

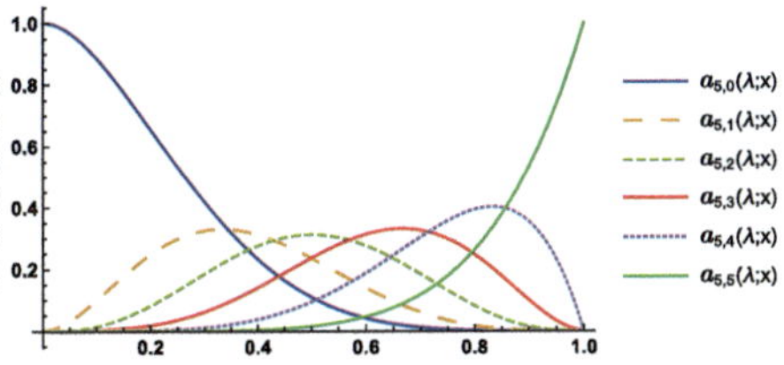


(D) $a_{5,i}(\lambda; x)$ with $\lambda_i = -\binom{5}{i}$, $i = 1, 2, 3$ and $\lambda_i = \binom{5}{i}$, $i = 4, 5$

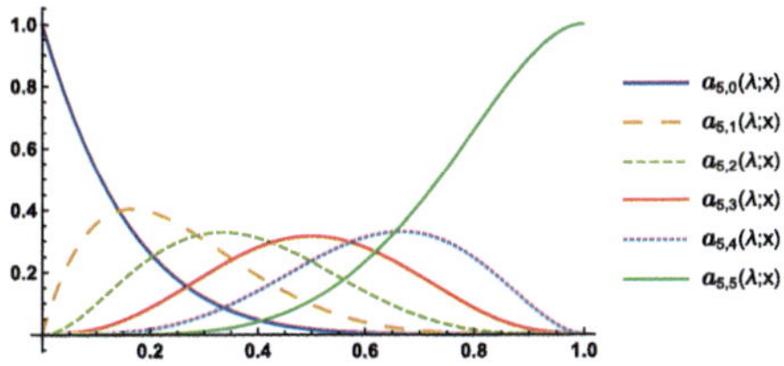


(E) $a_{5,i}(\lambda; x)$ with $\lambda_i = \binom{5}{i-1}$, $i = 1, 2, 3$ and $\lambda_i = -\binom{5}{i-1}$, $i = 4, 5$

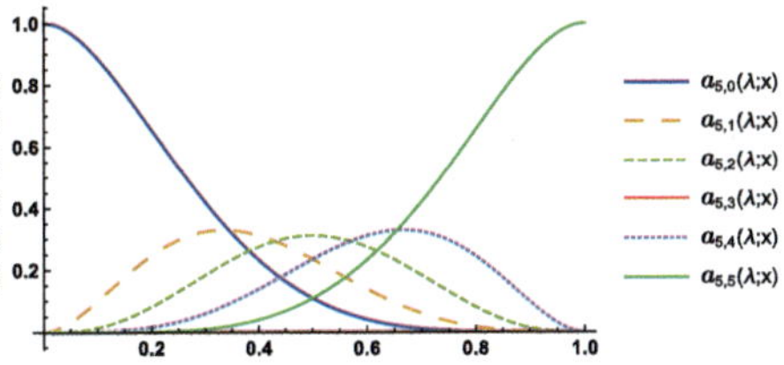


(F) $a_{5,i}(\lambda; x)$ with $\lambda_i = -\binom{5}{i}$, $i = 1, 2, 3$ and $\lambda_i = -\binom{5}{i-1}$, $i = 4, 5$

**Fig. 1** Behavior of polynomials $b_{n,i}(x)$ and $a_{n,i}(\lambda; x)$ for $n = 5$ and $i = 0, 1, \ldots, 5$

**Example 6.1** Consider the function

$$f(x) = \begin{cases} 8x & 0 \leq x \leq \frac{1}{5} \\ \frac{4(1+x)}{3} & \frac{1}{5} < x \leq \frac{1}{2} \\ \frac{4(2-x)}{3} & \frac{1}{2} < x \leq \frac{4}{5} \\ 8(1-x) & \frac{4}{5} < x \leq 1 \end{cases}$$

on the interval [0, 1] (see [17]). Behavior of approximation properties of $B_n(f; x; \lambda)$ and related error of approximations are given in Fig. 2 (A–C). A comparison between convergence of the classical Bernstein operators $B_n(f; x)$ and generalized Bernstein operators $B_n(f; x; \lambda)$ to $f(x)$ is shown in part (B) of Fig. 2 for $n = 10$ and $\lambda_i = \binom{n}{i-1}$, $i = 1, 2, \ldots, [\frac{n}{2}]$, $\lambda_i = \binom{n}{i}$, $i = [\frac{n}{2}] + 1, \ldots, n$. It can be seen that the error estimation obtained from modified operators is better than that of the classical Bernstein operators for the function $f(x)$.

**Example 6.2** Consider the function

$$g(x) = x sin(3\pi x)$$

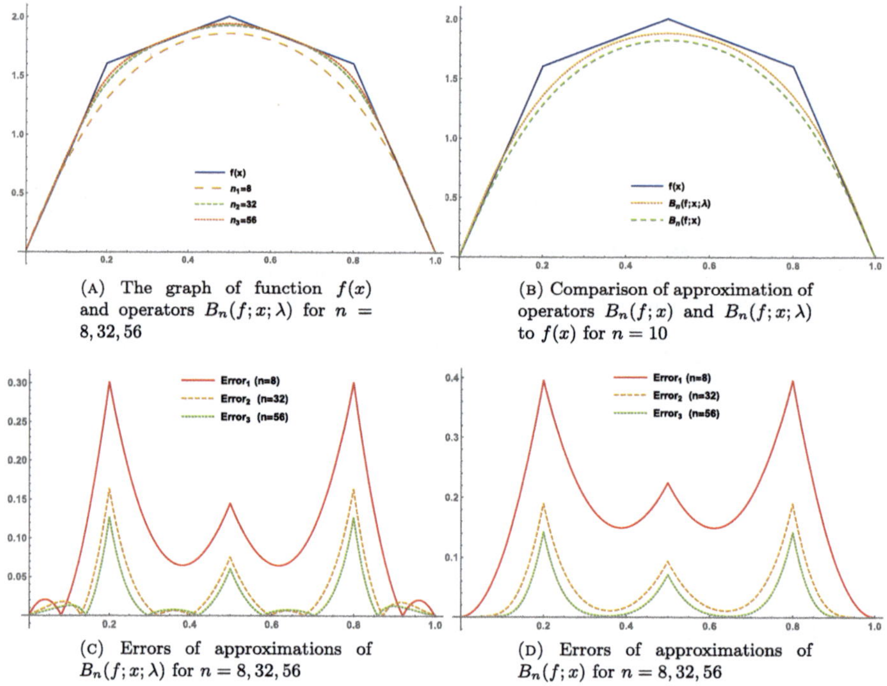


(A) The graph of function $f(x)$ and operators $B_n(f; x; \lambda)$ for $n = 8, 32, 56$

(B) Comparison of approximation of operators $B_n(f; x)$ and $B_n(f; x; \lambda)$ to $f(x)$ for $n = 10$

(C) Errors of approximations of $B_n(f; x; \lambda)$ for $n = 8, 32, 56$

(D) Errors of approximations of $B_n(f; x)$ for $n = 8, 32, 56$

**Fig. 2** Behavior of approximation properties of $B_n(f; x; \lambda)$ and $B_n(f; x)$ and related error of approximations

on the interval [0, 1]. In Fig. 3, we study approximation behavior of polynomials $B_n(g; x; \lambda)$ and related error of approximations for different values of $\lambda$. It is clear from the graphs that $B_n(g; x; \lambda)$ gives a better approximation on [0, 1].

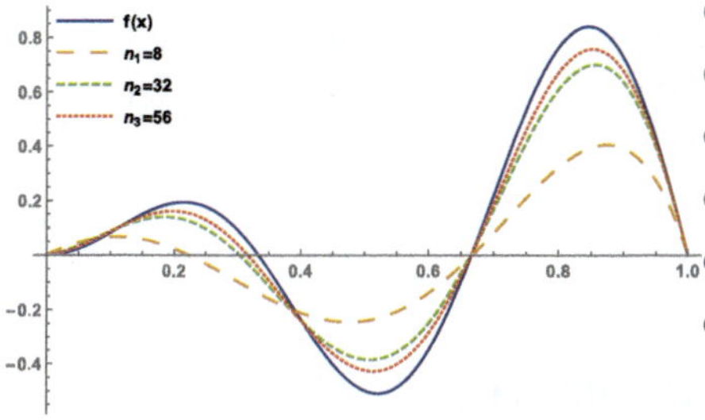


(A) Approximation of operators $B_n(g;x;\lambda)$ with $\lambda_i = -\binom{n}{i}$, $i = 1, 2, \cdots, \frac{n}{2}$ and $\lambda_i = \binom{n}{i}$, $i = \frac{n}{2} + 1, \cdots, n$

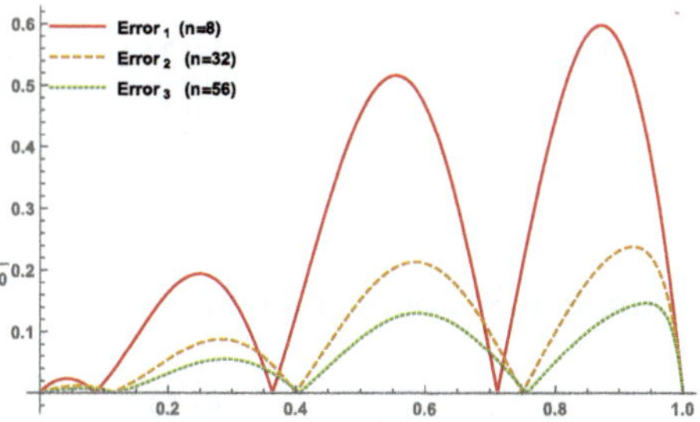


(B) Error of approximation of operators $B_n(g;x;\lambda)$ with $\lambda_i = -\binom{n}{i-1}$, $i = 1, 2, \cdots, \frac{n}{2}$ and $\lambda_i = \binom{n}{i-1}$, $i = \frac{n}{2} + 1, \cdots, n$

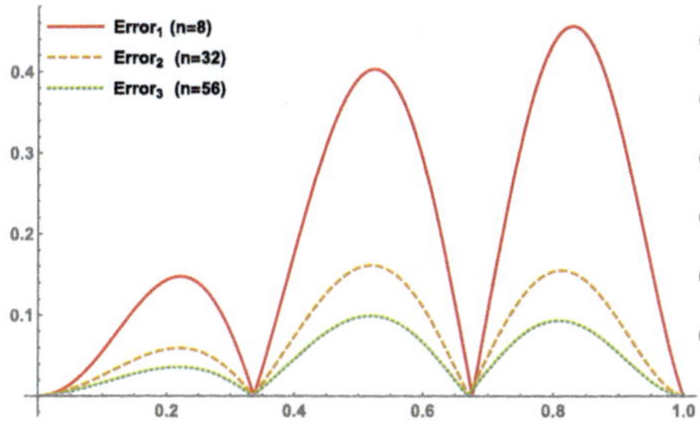


(C) Error of approximation of operators $B_n(g;x;\lambda)$ with $\lambda_i = -\binom{n}{i}$, $i = 1, 2, \cdots, \frac{n}{2}$ and $\lambda_i = \binom{n}{i}$, $i = \frac{n}{2} + 1, \cdots, n$

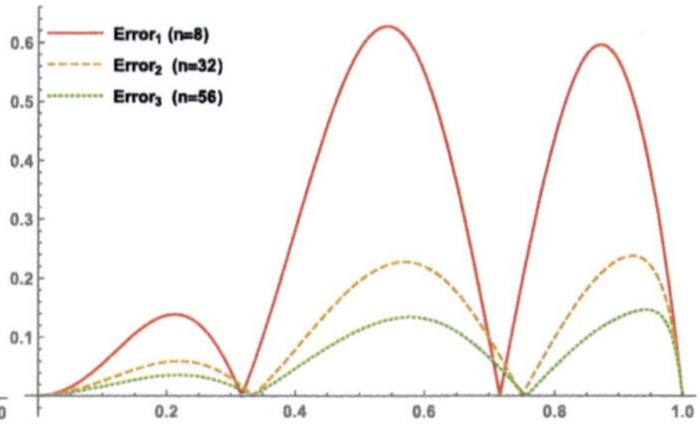


(D) Error of approximation of operators $B_n(g;x;\lambda)$ with $\lambda_i = -\binom{n}{i}$, $i = 1, 2, \cdots, \frac{n}{2}$ and $\lambda_i = -\binom{n}{i-1}$, $i = \frac{n}{2} + 1, \cdots, n$

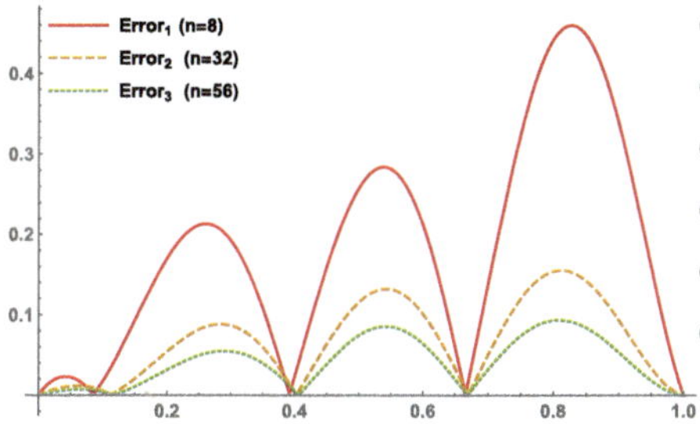


(E) Error of approximation of operators $B_n(g;x;\lambda)$ with $\lambda_i = \binom{n}{i-1}$, $i = 1, 2, \cdots, \frac{n}{2}$ and $\lambda_i = \binom{n}{i}$, $i = \frac{n}{2} + 1, \cdots, n$

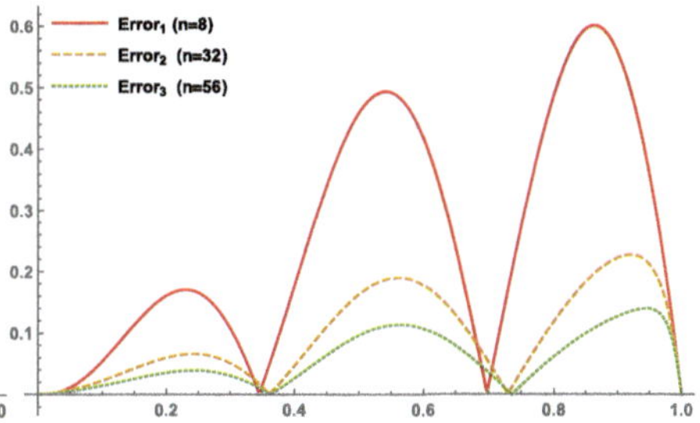


(F) Error of approximation of operators $B_n(g;x;\lambda)$ with $\lambda_i = \binom{n}{i-1} - \binom{n}{i}$, $i = 1, 2, \cdots, \frac{n}{2}$ and $\lambda_i = \binom{n}{i} - \binom{n}{i-1}$, $i = \frac{n}{2} + 1, \cdots, n$

**Fig. 3** Behavior of approximation of operators $B_n(g; x; \lambda)$ and related error of approximations for certain shape parameters $\lambda$

**Table 1** Maximum error of approximation

| $h(x) = \|1 - x\| sin(2\pi x)$ | $n = 4$ | $n = 6$ | $n = 8$ | $n = 10$ | $n = 12$ | $n = 14$ |
|---|---|---|---|---|---|---|
| $\\|f - B_n(h; \lambda_a)\\|_\infty$ | 0.394 | 0.285 | 0.234 | 0.203 | 0.181 | 0.164 |
| $\\|f - B_n(h; \lambda_b)\\|_\infty$ | 0.456 | 0.338 | 0.271 | 0.227 | 0.195 | 0.171 |
| $\\|f - B_n(h; \lambda_c)\\|_\infty$ | 0.462 | 0.343 | 0.274 | 0.229 | 0.196 | 0.172 |
| $\\|f - B_n(h; \lambda_d)\\|_\infty$ | 0.469 | 0.349 | 0.280 | 0.233 | 0.200 | 0.175 |
| $\\|f - B_n(h)\\|_\infty$ | 0.461 | 0.342 | 0.274 | 0.228 | 0.196 | 0.172 |

**Example 6.3** Consider the function

$$h(x) = |1 - x|\, sin(2\pi x)$$

on the interval [0, 1]. Table 1 represents the maximum error of approximation of our operators for the following shape parameters $\lambda$:

$$\lambda_a = \begin{cases} \lambda_i = 0.4\left(\binom{n}{i-1} + \binom{n}{i}\right), & i = 1, 2, \ldots, [\frac{n}{2}] \\ \lambda_i = 0.99\left(\binom{n}{i} + \binom{n}{i-1}\right), & i = [\frac{n}{2}] + 1, \ldots, n \end{cases} \quad \lambda_b = \begin{cases} \lambda_i = 0.01\binom{n}{i-1}, & i = 1, 2, \ldots, [\frac{n}{2}] \\ \lambda_i = \binom{n}{i}, & i = [\frac{n}{2}] + 1, \ldots, n \end{cases}$$

$$\lambda_c = \begin{cases} \lambda_i = 0.01\binom{n}{i-1}, & i = 1, 2, \ldots, [\frac{n}{2}] \\ \lambda_i = -0.09\binom{n}{i-1}, & i = [\frac{n}{2}] + 1, \ldots, n \end{cases} \quad \lambda_d = \begin{cases} \lambda_i = -0.05\binom{n}{i}, & i = 1, 2, \ldots, [\frac{n}{2}] \\ \lambda_i = -0.01\binom{n}{i-1}, & i = [\frac{n}{2}] + 1, \ldots, n \end{cases}$$

We also compare the maximum error of our operators and the classical Bernstein operators in this table.

The results that we obtain in this section not only support our theoretical results given in previous sections but also give more accurate results compared with Bersntein type operators and their modified forms.

## References

1. V.K. Weierstrass, Ueber die analytische Darstellbarkeit sogennanter willkürlicher Functionen einer reellen Veranderlichen, in: *Sitzungsberichte der Akademie zu Berlin*, pp. 633–639, 789–805 (1885)
2. S.N. Bernstein, constructive proof of Weierstrass approximation theorem. Commun. Kharkov Math. Soc. (1912)
3. G.G. Lorentz, *Bernstein Polynomials* (University of Toronto Press, Toronto, 1953)
4. H. Oruç, G.M. Phillips, A generalization of Bernstein polynomials Proc. Edinburgh Math. Soc. **42**, 403–413 (1999)
5. A. Ilinskii, S. Ostrovska, Convergence of generalized Bernstein polynomials. J. Approx. Theory **116**(1), 100–112 (2002)
6. M. Ayman Mursaleen, A. Kılıçman, Md. Nasiruzzaman, Approximation by q-Bernstein-Stancu-Kantorovich operators with shifted knots of real parameters. Filomat **36**(4), 1179–1194 (2022)

7. H.M. Srivastava, K.J. Ansari, F. Özger, Z. Ödemis Özger, A link between approximation theory and summability methods via four-dimensional infinite matrices. Mathematics *9*, 1895 (2021)
8. K.J. Ansari, F. Özger, Z. Ödemis Özger Numerical and theoretical approximation results for Schurer-Stancu operators with shape parameter $\lambda$. Comput. Appl. Math. *41*, 1–18 (2022)
9. X. Chen, J. Tan, Z. Liu, J. Xie, Approximation of functions by a new family of generalized Bernstein operators. J. Math. Anal. Appl. **450**, 244–261 (2017)
10. S.A. Mohiuddine, F. Özger, Approximation of functions by Stancu variant of Bernstein-Kantorovich operators based on shape parameter $\alpha$. RACSAM **114**, 70 (2020)
11. H.M. Srivastava, F. Özger, S.A. Mohiuddine, Construction of Stancu-type Bernstein operators based on Bézier bases with shape parameter $\lambda$. Symmetry **11**(3), Article 316 (2019)
12. F. Özger, Weighted statistical approximation properties of univariate and bivariate $\lambda$-Kantorovich operators. Filomat **33**(11), 473–3486 (2019)
13. F. Özger, On new Bézier bases with Schurer polynomials and corresponding results in approximation theory. Commun. Fac. Sci. Univ. Ank. Ser. A1 Math. Stat. **69**(1), 376–393 (2020)
14. G. Hu, C. Bo, X. Qin, Continuity conditions for $q$-Bezier curves of degree n. J. Inequal. Appl. **2017**, 115 (2017)
15. K. Khan, D.K. Lobiyal, Bézier curves based on Lupaş (p, q)-analogue of Bernstein functions in CAGD. J. Comput. Appl. Math. **317**, 458–477 (2017)
16. X. Qin, G. Hu, N. Zhang, X. Shen, Y. Yang, A novel extension to the polynomial basis functions describing Bezier curves and surfaces of degree n with multiple shape parameters. Appl. Math. Comput. **223**, 1–16 (2013)
17. H. Khosravian-Arab, M. Dehghan, M.R. Eslahchi, A new approach to improve the order of approximation of the Bernstein operators: theory and applications. Numer. Algor. **77**, 111–150 (2018)
18. R. Goldman, X. Xu, X. Zeng, Applications of the Shorgin Identity to Bernstein type operators. Res. Math. **73**, 2 (2018)
19. S. Deniz, F. Özger, Z. Ödemis Özger, S.A. Mohiuddine, M. Temizer Ersoy, Numerical solution of fractional Volterra integral equations based on rational Chebyshev approximation. Miskolc Math. Notes **24**(3), 1287–1305 (2023)
20. R. Aslan, Some approximation results on $\lambda$-Szász-Mirakjan-Kantorovich operators. Fundam. J. Math. Appl. **4**(3), 150–158 (2021)
21. X. Han, Y.C. Ma, X.L. Huang, A novel generalization of Bezier curve and surface. J. Comput. Appl. Math. **217**, 180–193 (2008)
22. H. Bohman, On approximation of continuous and analytic functions. Ark. Math. **2**, 43–56 (1952)
23. P.P. Korovkin, On convergence of linear positive operators in the space of continuous. Dokl. Akad. Nauk. SSSR (N.S) **90**, 961–964 (1953)
24. J. Peetre, Theory of interpolation of normed spaces. Notas Mat. Rio de Janeiro **39**, 1–86 (1963)
25. R.A. DeVore, G.G. Lorentz, *Constructive Approximation* (Springer, Berlin, 1993)
26. Z. Ditzian, V. Totik, *Moduli of Smoothness* (Springer, New York, 1987)
27. M.A. Ozarslan, H. Aktuglu, Local approximation for certain King type operators. Filomat **27**(1), 173–181 (2013)
28. A.D. Gadjiev, C. Orhan, Some approximation theorems by statistical convergence. Rocky Mountain J. Math. **32**, 129–138 (2002)
29. Md. Nasiruzzaman, A. Kilicman, M. Ayman Mursaleen, Construction of $q$-Baskakov operators by wavelets and approximation properties. Iranian J. Sci. Technol. Trans. A: Sci. **46**(5), 1495–1503 (2022)
30. U. Kadak, Generalized statistical convergence based on fractional order difference operator and its applications to approximation theorems. Iran. J. Sci. Technol. Trans. Sci. **43**, 225–237 (2019)
31. U. Kadak, Generalized weighted invariant mean based on fractional difference operator with applications to approximation theorems for functions of two variables. Res. Math. **72**, 1181–1202 (2017)

32. P. Baliarsingh, U. Kadak, M. Mursaleen, On statistical convergence of difference sequences of fractional order and related Korovkin type approximation theorems. Quaestiones Mathematicae **41**(8), 1117–1133 (2018)
33. H. Fast, Sur la convergence statistique. Colloq. Math. **2**, 241–244 (1951)
34. E. Kolk, Matrix summability of statistically convergent sequences. Analysis **13**, 77–83 (1993)
35. U. Kadak, Modularly weighted four dimensional matrix summability with application to Korovkin type approximation theorem. J. Math. Anal. Appl. **468**(1), 38–63 (2018)
36. U. Kadak, N.L. Braha, H.M. Srivastava, Statistical weighted B-summability and its applications to approximation theo-rems. Appl. Math. Comput. **302**, 80–96 (2017)
37. U. Kadak, Weighted statistical relative invariant mean in modular function spaces with related approximation results. Numer. Funct. Anal. Optim. **39**(11), 1181–1207
38. S. A. Mohiuddine, Statistical weighted $A$-summability with application to Korovkins type approximation theorem. J. Inequal. Appl. **2016**, Article ID 101 (2016)
39. R. Aslan, Rate of approximation of blending type modified univariand bivariate $\lambda$ -Schurer-Kantorovich operators. Kuwait. J. Sci. (2024). https://doi.org/10.1016/j.kjs.2023.12.007
40. Z. Finta, Remark on Voronovskaja theorem for q-Bernstein operators. Stud. Univ. Babes-Bolyai, Math. **56**, 335–339 (2011)

# Approximation by $q$-analogue of Modified Szász Operators Generated by Dunkl Exponential Functions

**Md. Nasiruzzaman and S. A. Mohiuddine**

**Abstract** We define a modification of Szász operators generated by Dunkl exponential functions via $q$-calculus based on a continuously differentiable function $\tau$ on $[0, \infty)$. The uniform approximation and weighted approximation of aforesaid operators are obtained. Moreover, the rate of convergence of our operators for functions belonging to the Lipschitz class including some other related approximation results is presented.

**Keywords** $q$-integers · Dunkl analogue · Szász operator · Modulus of continuity · Peetre's $K$-functional

**AMS Subject Classification (2010)** 41A25 · 41A36 · 33C45

## 1 Introduction and Preliminaries

In 1912, Bernstein [7] introduced the following sequence of positive linear operators $B_n : C[0, 1] \to C[0, 1]$ defined by

$$B_n(f; x) = \sum_{k=0}^{n} \binom{n}{k} x^k (1-x)^{n-k} f\left(\frac{k}{n}\right), \qquad x \in [0, 1] \tag{1.1}$$

for $n \in \mathbb{N}$ and $f \in C[0, 1]$.

Md. Nasiruzzaman
Department of Mathematics, Faculty of Science, University of Tabuk, PO Box-4279, Tabuk 71491, Saudi Arabia
e-mail: nasir3489@gmail.com; mfarooq@ut.edu.sa

S. A. Mohiuddine (✉)
Department of General Required Courses, Mathematics, The Applied College, King Abdulaziz University, Jeddah, Saudi Arabia
e-mail: mohiuddine@gmail.com

Operator Theory and Applications Research Group, Department of Mathematics, Faculty of Science, King Abdulaziz University, Jeddah, Saudi Arabia

 
S. A. Mohiuddine et al. (eds.), *Positive Operators and Fixed-Point Theorems with Applications*, Industrial and Applied Mathematics,
https://doi.org/10.1007/978-981-96-2476-8_7

In 1950, for $x \geq 0$, Szász [36] introduced the operators

$$S_n(f;x) = e^{-nx}\sum_{k=0}^{\infty}\frac{(nx)^k}{k!}f\left(\frac{k}{n}\right), \qquad f \in C[0,\infty). \tag{1.2}$$

In the field of approximation theory, the application of $q$-calculus emerged as a new area to study approximation properties of positive linear operators. The first $q$-analogue of the well-known Bernstein polynomials was introduced by Lupaş [19]. Later, Phillips [31] considered another $q$-analogue of the classical Bernstein polynomials. We now present some basic definitions and details on $q$-calculus which will be used in this paper (see [16]).

For $0 < q \leq 1$ and $\lambda \in \mathbb{N}_0$, the $q$-integer defined by

$$[\lambda]_q = \frac{1-q^\lambda}{1-q}, \qquad q \neq 1$$

and $[\lambda]_q = \lambda$ for $q = 1$. The $q$-factorial is $[\lambda]_q! = [\lambda]_q\,[\lambda-1]_q \ldots [1]_q$ for $\lambda \in \mathbb{N}$ and $[\lambda]_q! = 1$ for $\lambda = 0$. For $1 \leq n \leq \lambda$ and $\lambda \in \mathbb{N}$, the $q$-binomial coefficient

$$\begin{bmatrix}\lambda\\ n\end{bmatrix}_q = \frac{[\lambda]_q!}{[\lambda-n]_q!\,[n]_q!}$$

with

$$\begin{bmatrix}\lambda\\ 0\end{bmatrix}_q = 1 \qquad \text{and} \qquad \begin{bmatrix}\lambda\\ n\end{bmatrix}_q = 0 \quad for \quad n > \lambda.$$

Also

$$(a;q)_\lambda := \begin{cases} 1 & (\lambda = 0) \\ \prod\limits_{n=0}^{\lambda-1}(1-aq^n) & (\lambda \in \mathbb{N}) \end{cases}$$

and

$$(a;q)_\infty := \prod_{n=0}^{\infty}\left(1-aq^n\right).$$

Recall the following generalized exponential functions (see [32]) defined by

$$e_\mu(x) = \sum_{n=0}^{\infty}\frac{x^n}{\gamma_\mu(n)}.$$

Here

$$\gamma_\mu(2k) = \frac{2^{2k}k!\,\Gamma\left(k+\mu+\frac{1}{2}\right)}{\Gamma\left(\mu+\frac{1}{2}\right)},$$

and

$$\gamma_\mu(2k+1) = \frac{2^{2k+1}k!\Gamma\left(k+\mu+\frac{3}{2}\right)}{\Gamma\left(\mu+\frac{1}{2}\right)}.$$

The recursion formula for $\gamma_\mu$ is given by

$$\gamma_\mu(k+1) = (k+1+2\mu\theta_{k+1})\gamma_\mu(k), \quad k = 0, 1, 2, \ldots,$$

where $\mu > -\frac{1}{2}$ and

$$\theta_k = \begin{cases} 0 & \text{if } k \in 2\mathbb{N} \\ 1 & \text{if } k \in 2\mathbb{N}+1. \end{cases}$$

Sucu [35] constructed a Dunkl analogue of Szász operators with the help of generalized exponential functions defined in [32] as follows:

$$S_n^*(f; x) := \frac{1}{e_\mu(nx)} \sum_{k=0}^{\infty} \frac{(nx)^k}{\gamma_\mu(k)} f\left(\frac{k+2\mu\theta_k}{n}\right), \tag{1.3}$$

where $x \geq 0, \quad f \in C[0, \infty), \mu \geq 0, \quad n \in \mathbb{N}$.

Cheikh et al. [9] defined the notions of $q$-Dunkl analogues of exponential functions and recursion relations for $\mu > -\frac{1}{2}$ and $0 < q < 1$ as

$$e_{\mu,q}(x) = \sum_{n=0}^{\infty} \frac{x^n}{\gamma_{\mu,q}(n)}, \quad x \in [0, \infty) \tag{1.4}$$

$$E_{\mu,q}(x) = \sum_{n=0}^{\infty} \frac{q^{\frac{n(n-1)}{2}} x^n}{\gamma_{\mu,q}(n)}, \quad x \in [0, \infty) \tag{1.5}$$

$$\gamma_{\mu,q}(n+1) = \left(\frac{1-q^{2\mu\theta_{n+1}+n+1}}{1-q}\right)\gamma_{\mu,q}(n), \quad n \in \mathbb{N}, \tag{1.6}$$

$$\theta_n = \begin{cases} 0 & \text{if } n \in 2\mathbb{N}, \\ 1 & \text{if } n \in 2\mathbb{N}+1. \end{cases}$$

An explicit formula for $\gamma_{\mu,q}(n)$ is

$$\gamma_{\mu,q}(n) = \frac{(q^{2\mu+1}, q^2)_{[\frac{n+1}{2}]}(q^2, q^2)_{[\frac{n}{2}]}}{(1-q)^n}\gamma_{\mu,q}(n), \quad n \in \mathbb{N}.$$

Some special cases of $\gamma_{\mu,q}(n)$ are as follows:

$$\gamma_{\mu,q}(0) = 1, \qquad \gamma_{\mu,q}(1) = \frac{1-q^{2\mu+1}}{1-q}$$

$$\gamma_{\mu,q}(2) = \left(\frac{1-q^{2\mu+1}}{1-q}\right)\left(\frac{1-q^{2}}{1-q}\right),$$

$$\gamma_{\mu,q}(3) = \left(\frac{1-q^{2\mu+1}}{1-q}\right)\left(\frac{1-q^{2}}{1-q}\right)\left(\frac{1-q^{2\mu+3}}{1-q}\right),$$

$$\gamma_{\mu,q}(4) = \left(\frac{1-q^{2\mu+1}}{1-q}\right)\left(\frac{1-q^{2}}{1-q}\right)\left(\frac{1-q^{2\mu+3}}{1-q}\right)\left(\frac{1-q^{4}}{1-q}\right).$$

In [15], Içöz and Çekim gave the Dunkl generalization of Szász operators via $q$ -calculus as

$$D_{n,q}(f;x) = \frac{1}{e_{\mu,q}([n]_q x)} \sum_{k=0}^{\infty} \frac{([n]_q x)^k}{\gamma_{\mu,q}(k)} f\left(\frac{1-q^{2\mu\theta_k+k}}{1-q^n}\right), \tag{1.7}$$

for $\mu > \frac{1}{2}$, $x \geq 0$, $0 < q < 1$ and $f \in C[0,\infty)$.

Previous studies demonstrate that better error estimation for positive linear operators plays an important role in approximation theory which allows us to approximate much faster to the function being approximated. In [11, 12, 25], various approximation properties of Szász–Mirakjan–Kantorovich operators and their quantum variant were investigated.

In this chapter, we construct the $q$-analogue of Szász operators generated by Dunkl exponential functions based on a continuously differentiable function $\tau$ on $[0,\infty)$ while generalized Szász–Mirakyan operators based on $\tau$ were recently studied in [4] (also see [8, 33]) and then we study the uniform approximation, degree of approximation as well as several related approximation results. We refer to [1–3, 5, 6, 20–24, 26–30, 34, 37] for more details on approximation by positive linear operators.

## 2 Construction of Operators and Auxiliary Lemmas

To define our operators, we first consider that $\tau(x)$ is a continuously differentiable functions defined on $\mathbb{R}^+ = [0,\infty)$ satisfying

(1) $\tau(0) = 0$,
(2) $\inf_{x\in\mathbb{R}^+} \tau'(x) \geq 1$.

For any $0 < q < 1$, $\mu > \frac{1}{2n}$, and $n \in \mathbb{N}$, we define $q$-analogue of Szász operators generated by Dunkl exponential functions based on the above function $\tau$ by

$$
\begin{aligned}
L_{n,q}^{\tau}(f;x) &= \frac{1}{e_{\mu,q}([n]_q\tau(x))}\sum_{k=0}^{\infty}\frac{([n]_q\tau(x))^k}{\gamma_{\mu,q}(k)}(fo\tau^{-1})\left(\frac{1-q^{2\mu\theta_k+k}}{1-q^n}\right)\\
&= (D_{n,q}(fo\tau^{-1})o\tau)(x)\\
&= \frac{1}{e_{\mu,q}([n]_q\tau(x))}\sum_{k=0}^{\infty}\frac{([n]_q\tau(x))^k}{\gamma_{\mu,q}(k)}f\left(\tau^{-1}\left(\frac{1-q^{2\mu\theta_k+k}}{1-q^n}\right)\right), \quad (2.1)
\end{aligned}
$$

where $e_{\mu,q}(x)$ and $\gamma_{\mu,q}$ are defined in (1.4) and (1.6), respectively, and $f \in C_\zeta[0,\infty)$ with $\zeta \geq 0$ and

$$
C_\zeta[0,\infty) = \{f \in C[0,\infty) :\mid f(t) \mid\leq M(1+t)^\zeta, \quad \text{for some} \quad M>0, \quad \zeta>0\}. \quad (2.2)
$$

Note that the function $\tau(x) = x + x^2$ satisfies the conditions 1 and 2. If $\tau = t$, then $L_{n,q}^{\tau} = D_{n,q}$.

**Lemma 2.1** *Let $L_{n,q}^{\tau}(.\ ;\ .)$ be the operators given by (2.1). Then, for each continuously differentiable function $\tau$ on $\mathbb{R}^+$, we have the following identities:*

*(1)* $L_{n,q}^{\tau}(1;x) = 1$,
*(2)* $L_{n,q}^{\tau}(\tau;x) = \tau(x)$,
*(3)* $\tau(x)^2 + \left(q^{2\mu}[1-2\mu]_q\frac{e_{\mu,q}(q[n]_q\tau(x))}{e_{\mu,q}([n]_q\tau(x))}\right)\frac{\tau(x)}{[n]_q} \leq L_{n,q}^{\tau}(\tau^2;x) \leq \tau(x)^2 + ([1+2\mu]_q)\frac{\tau(x)}{[n]_q}$.

***Proof*** (1) It is to see that

$$
L_{n,q}^{\tau}(1;x) = \frac{1}{e_{\mu,q}([n]_q\tau(x))}\sum_{k=0}^{\infty}\frac{([n]_q\tau(x))^k}{\gamma_\mu(k)} = 1.
$$

(2) We can write

$$
\begin{aligned}
L_{n,q}^{\tau}(\tau;x) &= \frac{1}{e_{\mu,q}([n]_q\tau(x))}\sum_{k=0}^{\infty}\frac{([n]_q\tau(x))^k}{\gamma_\mu(k)}\left(\frac{1-q^{2\mu\theta_k+k}}{1-q^n}\right)\\
&= \frac{1}{[n]_q e_{\mu,q}([n]_q\tau(x))}\sum_{k=1}^{\infty}\frac{([n]_q\tau(x))^k}{\gamma_\mu(k-1)}\\
&= \tau(x).
\end{aligned}
$$

(3) We can write

$$\begin{aligned} L_{n,q}^{\tau}(\tau^2; x) &= \frac{1}{e_{\mu,q}([n]_q\tau(x))} \sum_{k=0}^{\infty} \frac{([n]_q\tau(x))^k}{\gamma_\mu(k)} \left(\frac{1-q^{2\mu\theta_k+k}}{1-q^n}\right)^2 \\ &= \frac{1}{[n]_q^2 e_{\mu,q}([n]_q\tau(x))} \sum_{k=0}^{\infty} \frac{([n]_q\tau(x))^k}{\gamma_\mu(k-1)} \left(\frac{1-q^{2\mu\theta_k+k}}{1-q}\right) \\ &= \frac{1}{[n]_q^2 e_{\mu,q}([n]_q\tau(x))} \sum_{k=0}^{\infty} \frac{([n]_q\tau(x))^{k+1}}{\gamma_\mu(k)} \left(\frac{1-q^{2\mu\theta_{k+1}+k+1}}{1-q}\right). \end{aligned}$$

From [15], we know that

$$[2\mu\theta_{k+1} + k + 1]_q = [2\mu\theta_k + k]_q + q^{2\mu\theta_k+k}[2\mu(-1)^k + 1]_q. \tag{2.3}$$

Now, by separating the even and odd terms and using (2.3), we get

$$\begin{aligned} L_{n,q}^{\tau}(\tau^2; x) &= \frac{1}{[n]_q^2 e_{\mu,q}([n]_q\tau(x))} \sum_{k=0}^{\infty} \frac{([n]_q\tau(x))^{k+1}}{\gamma_\mu(k)} \left(\frac{1-q^{2\mu\theta_{k+1}+k+1}}{1-q}\right) \\ &+ \frac{[1+2\mu]_q}{[n]_q^2 e_{\mu,q}([n]_q\tau(x))} \sum_{k=0}^{\infty} \frac{([n]_q\tau(x))^{2k+1}}{\gamma_\mu(2k)} q^{2\mu\theta_{2k}+2k} \\ &+ \frac{[1-2\mu]_q}{[n]_q^2 e_{\mu,q}([n]_q\tau(x))} \sum_{k=0}^{\infty} \frac{([n]_q\tau(x))^{2k+2}}{\gamma_\mu(2k)} q^{2\mu\theta_{2k+1}+2k+1}. \end{aligned}$$

We know the inequality

$$[1 - 2\mu]_q \leq [1 + 2\mu]_q. \tag{2.4}$$

Therefore, by using (2.4), we have

$$\begin{aligned} L_{n,q}^{\tau}(\tau^2; x) &\geq (\tau(x))^2 + \frac{\tau(x)[1-2\mu]_q}{[n]_q e_{\mu,q}([n]_q\tau(x))} \sum_{k=0}^{\infty} \frac{(q[n]_q\tau(x))^{2k}}{\gamma_\mu(2k)} \\ &+ \frac{q^{2\mu}\tau(x)[1-2\mu]_q}{[n]_q e_{\mu,q}([n]_q\tau(x))} \sum_{k=0}^{\infty} \frac{(q[n]_q\tau(x))^{2k+1}}{\gamma_\mu(2k+1)} \\ &\geq (\tau(x))^2 + q^{2\mu}[1-2\mu]_q \frac{e_{\mu,q}(q[n]_q\tau(x))}{e_{\mu,q}([n]_q\tau(x))} \frac{\tau(x)}{[n]_q}. \end{aligned}$$

Similarly, on the other hand, we have

$$L_{n,q}^{\tau}(\tau^2; x) \leq (\tau(x))^2 + [1+2\mu]_q \frac{\tau(x)}{[n]_q}$$

which completes the proof. □

**Lemma 2.2** *Let the operators* $L^{\tau}_{n,q}(.\ ;\ .)$ *be given by* (2.1). *Then, for each continuously differentiable function* $\tau(x)$ *on* $\mathbb{R}^+$, *we have*

*(1)* $L^{\tau}_{n,q}(\tau - \tau(x); x) = 0,$

*(2)* $L^{\tau}_{n,q}((\tau - \tau(x))^2; x) \le [1 + 2\mu]_q \frac{\tau(x)}{[n]_q}.$

## 3 Weighted Approximation

In this section, we study the Korovkin-type approximation properties for our operators defined in (2.1).

Korovkin theory is well-established area of research in approximation theory. Korovkin [17] established the necessary and sufficient conditions for the uniform convergence of the sequence of positive linear operators on the space of real-valued continuous functions on a compact subset of real numbers to a function $f$ by using the test functions $e_i$ defined by $e_i = x_i$ $(i = 0, 1, 2)$. Gadžiev [13, 14] presented an extension of this approximation theorem to unbounded intervals as well as weighted Korovkin-type theorem in a subspace of continuous functions on the real axis $\mathbb{R}$. It was also demonstrated that, under some additional conditions on $\tau$, the above test functions can be replaced by $1, \tau, \tau^2$ (see Theorem 3.2 of this section).

Let $C_B(\mathbb{R}^+)$ be the set of all bounded and continuous functions on $\mathbb{R}^+ = [0, \infty)$. The function $f \in C_B(\mathbb{R}^+)$ is a linear normed space with

$$\| f \|_{C_B} = \sup_{x \ge 0} | f(x) | .$$

Let

$$H := \left\{ f : x \in [0, \infty), \frac{f(x)}{1 + \tau(x)^2} \text{ is convergent as } x \to \infty \right\}.$$

**Theorem 3.1** *Let* $L^{\tau}_{n,q}(.\ ;\ .)$ *be the operators defined by* (2.1). *Then for any function* $f \in C_{\zeta}[0, \infty) \cap H, \quad \zeta \ge 2,$

$$\lim_{n \to \infty} L^{\tau}_{n,q}(f; x) = f(x)$$

*is uniformly on each compact subset of* $[0, \infty)$, *where* $x \in \left[\frac{1}{2}, b\right), \quad b > \frac{1}{2}$.

***Proof*** The proof is based on Lemma 2.1 and well-known Korovkin's theorem regarding the convergence of a sequence of linear and positive operators, so it is enough to prove the condition

$$\lim_{n \to \infty} L^{\tau}_{n,q}(\tau^j; x) = \tau(x)^j, \quad j = 0, 1, 2, \quad \{\text{as } n \to \infty\}$$

uniformly on $[0, 1]$.

Clearly $\frac{1}{[n]_q} \to 0$ $(n \to \infty)$, we have

$$\lim_{n\to\infty} L_{n,q}^{\tau}(\tau; x) = \tau(x), \quad \lim_{n\to\infty} L_{n,q}^{\tau}(\tau^2; x) = \tau(x)^2$$

which completes the proof. □

We recall the weighted spaces of the functions on $\mathbb{R}^+$, which are defined as follows:

$$P_\rho(\mathbb{R}^+) = \left\{f :\mid f(x) \mid\leq M_f \rho(x)\right\},$$
$$Q_\rho(\mathbb{R}^+) = \left\{f : f \in P_\rho(\mathbb{R}^+) \cap C[0, \infty)\right\},$$
$$Q_\rho^k(\mathbb{R}^+) = \left\{f : f \in Q_\rho(\mathbb{R}^+) \text{ and } \lim_{x\to\infty} \frac{f(x)}{\rho(x)} = k (k \text{ is a constant})\right\},$$

where $\rho(x) = 1 + \tau(x)^2$ is a weighted function and $M_f$ is a constant depending only on $f$. Note that $Q_\rho(\mathbb{R}^+)$ is a normed space with the norm

$$\| f \|_\rho = \sup_{x\geq 0} \frac{\mid f(x) \mid}{\rho(x)}.$$

**Lemma 3.1** ([13, 14]) *The linear positive operators $L_n$, $n \geq 1$ act from $Q_\rho(\mathbb{R}^+) \to P_\rho(\mathbb{R}^+)$ if and only if*

$$\| L_n(\varphi; x) \| \leq K\varphi(x),$$

*where $\varphi(x) = 1 + x^2$, $x \in \mathbb{R}^+$ and $K$ is a positive constant.*

**Theorem 3.2** ([13, 14]) *Let $\{L_n\}_{n\geq 1}$ be a sequence of positive linear operators acting from $Q_\rho(\mathbb{R}^+) \to P_\rho(\mathbb{R}^+)$ and satisfying the condition*

$$\lim_{n\to\infty} \| L_n(\rho^\nu) - \rho^\nu \|_\varphi = 0, \quad \nu = 0, 1, 2.$$

*Then, for any function $f \in Q_\rho^k(\mathbb{R}^+)$, we have*

$$\lim_{n\to\infty} \| L_n(f; x) - f \|_\varphi = 0.$$

**Theorem 3.3** *Let $L_{n,q}^{\tau}(.; .)$ be the operators defined by (2.1). Then, for each function $f \in Q_\rho^k(\mathbb{R}^+)$ we have*

$$\lim_{n\to\infty} \| L_{n,q}^{\tau}(f; x) - f \|_\rho = 0.$$

***Proof*** From Lemma 2.1, Theorem 3.2, and for $\nu = 0$, the first condition is fulfilled, that is, we have

$$\lim_{n\to\infty} \| L^{\tau}_{n,q}(1;x) - 1 \|_{\rho} = 0.$$

Similarly, from Lemma 2.1, Theorem 3.2, and for $\nu = 1, 2$ we have that

$$\begin{aligned}\sup_{x\in[0,\infty)} \frac{| L^{\tau}_{n,q}(\tau;x) - \tau(x) |}{1+\tau(x)^2} &\leq \frac{1}{2[n]_q} \sup_{x\in[0,\infty)} \frac{1}{1+\tau(x)^2} \\ &= \frac{1}{2[n]_q}\end{aligned}$$

which imply that

$$\lim_{n\to\infty} \| L^{\tau}_{n,q}(\tau;x) - \tau(x) \|_{\rho} = 0.$$

Now,

$$\begin{aligned}\sup_{x\in[0,\infty)} \frac{| L^{\tau}_{n,q}(\tau^2;x) - \tau(x)^2 |}{1+\tau(x)^2} &\leq \frac{| [1+2\mu]_q - 1 |}{[n]_q} \sup_{x\in[0,\infty)} \frac{\tau(x)}{1+\tau(x)^2} \\ &+ \frac{1}{4[n]_q^2} | [1+2\mu]_q - 1 | \sup_{x\in[0,\infty)} \frac{1}{1+\tau(x)^2}\end{aligned}$$

which imply that

$$\lim_{n\to\infty} \| L^{\tau}_{n,q}(\tau^2;x) - \tau(x)^2 \|_{\rho} = 0.$$

This completes the proof. □

## 4 Degree of Convergence

Here, we calculate the rate of convergence of our operators (2.1) by means of modulus of continuity and Lipschitz-type maximal functions.

Consider $\tau(x)$ is a continuously differentiable function on $\mathbb{R}^+$ and $x \geq \frac{1}{2n}$, $n \in \mathbb{N}$. For $\delta > 0$, the modulus of continuity of $f$ denoted by $\omega(f, \delta)$ gives the maximum oscillation of $f$ in any interval of length not exceeding $\delta > 0$ and it is given by

$$\omega(f,\delta) = \sup_{|\tau(t)-\tau(x)|\leq\delta} | f(t) - f(x) |, \quad t \in [0,\infty). \tag{4.1}$$

It is known that $\lim_{\delta\to 0+} \omega(f,\delta) = 0$ for $f \in C_B[0,\infty)$ and for any $\delta > 0$, one has

$$| f(t) - f(x) | \leq \left( \frac{| \tau(t) - \tau(x) |}{\delta} + 1 \right) \omega(f,\delta). \tag{4.2}$$

**Theorem 4.1** *Let $L^{\tau}_{n,q}(. \; ; \; .)$ be the operators defined by* (2.1). *Then, for $f \in C_B[0,\infty)$, we have*

$$\mid L_{n,q}^{\tau}(f;x) - f \mid \leq 2\omega\left(f;\delta_{n,\tau(x)}\right),$$

*where* $C_B[0,\infty)$ *is the space of uniformly continuous bounded functions on* $\mathbb{R}^+$, $\omega(f,\delta)$ *is the modulus of continuity of the function* $f \in C_B[0,\infty)$ *defined in* (4.1) *and*

$$\delta_{n,\tau(x)} = \sqrt{[1+2\mu]_q \frac{\tau(x)}{[n]_q}}. \tag{4.3}$$

***Proof*** We prove it by using (4.1), (4.2), and Cauchy–Schwarz inequality. We can easily get

$$\mid L_{n,q}^{\tau}(f;x) - f \mid \leq \left\{1 + \frac{1}{\delta}\left(L_{n,q}^{\tau}(\tau(t)-\tau(x))^2;x\right)^{\frac{1}{2}}\right\}\omega(f;\delta).$$

If we choose $\delta = \delta_{n,\tau(x)}$ and by applying the result (2) of Lemma 2.2, completes the proof. □

**Remark 4.1** For the operators $D_{n,q}(.\,;\ .)$ defined by (1.7), we may write, for every $f \in C_B[0,\infty)$, $x \geq 0$, and $n \in \mathbb{N}$, that

$$\mid D_{n,q}(f;x) - f(x) \mid \leq 2\omega\left(f;\lambda_{n,x}\right), \tag{4.4}$$

where, by [15], we have

$$\lambda_{n,x} = \sqrt{D_{n,q}((t-x)^2;x)} \leq \sqrt{[1+2\mu]_q \frac{x}{[n]_q}}. \tag{4.5}$$

Now, we claim that the error estimation in Theorem 4.1 is a generalization of (4.4) provided $f \in C_B[0,\infty)$ and $\tau(x)$ is a continuously differentiable function on $\mathbb{R}^+$. Indeed, for $\tau(x) = x$ and $x \in \mathbb{R}^+$, under these conditions, we notice that $\mid x - t \mid \leq \mid \tau(t) - \tau(x) \mid$, for every $x, t \in \mathbb{R}^+$. It is guaranteed that

$$L_{n,q}^{\tau}((\tau(t)-\tau(x))^2;x) = D_{n,q}((t-x)^2;x). \tag{4.6}$$

Next, we give the rate of convergence of our operators $L_{n,q}^{\tau}(f;x)$ in terms of the elements of the usual Lipschitz class $Lip_M(\nu)$.

Let $f \in C_B[0,\infty)$ and $0 < \nu \leq 1$. In [3], the Lipschitz-type maximal function space on $E \subset \mathbb{R}_+$ is defined as

$$\widetilde{W}_{\alpha,E} = \left\{f : \sup(1+\tau(x))^{\alpha}\widetilde{f}_{\alpha}(x) \leq M\frac{1}{(1+\tau(y))^{\alpha}} : \text{and } y \in E\right\}, \tag{4.7}$$

where $M$ is a positive constant, $0 < \alpha \leq 1$ and $\tau(x)$ continuously differentiable function on $\mathbb{R}^+$.

In [18], Lenze introduced a Lipschitz-type maximal function $f_{\alpha}$ as follows:

$$f_{\alpha}(x,t)=\sum_{\substack{t>0\\ t\neq x}}\frac{\mid f(t)-f(x)\mid}{\mid \tau(x)-\tau(t)\mid^{\alpha}}. \tag{4.8}$$

**Theorem 4.2** *Let $L^{\tau}_{n,q}(.\,;\ .)$ be the operator defined in* (2.1). *Then, for each $f\in Lip_M(\nu)$ $(M>0,\ 0<\nu\leq 1)$ satisfying* (4.7) *and* (4.8), *we have*

$$\mid L^{\tau}_{n,q}(f;x)-f\mid\leq M\left(\delta_{n,\tau(x)}\right)^{\frac{\nu}{2}},$$

*where $\delta_{n,\tau(x)}$ is given in Theorem 4.1.*

***Proof*** We prove it by using (4.7), (4.8), and Hölder inequality. We can write

$$\begin{aligned}\mid L^{\tau}_{n,q}(f;x)-f\mid &\leq \mid L^{\tau}_{n,q}(f(t)-f;x)\mid\\ &\leq L^{\tau}_{n,q}\left(\mid f(t)-f\mid;x\right)\\ &\leq \mid ML^{\tau}_{n,q}\left(\mid \tau(t)-\tau(x)\mid^{\nu};x\right).\end{aligned}$$

Therefore

$$\begin{aligned}&\mid L^{\tau}_{n,q}(f;x)-f\mid\\ &\leq M\frac{[n]_q}{e_{\mu,q}([n]_q\tau(x))}\sum_{k=0}^{\infty}\frac{([n]_q\tau(x))^k}{\gamma_{\mu,q}(k)}\left|\frac{1-q^{2\mu\theta_k+k}}{1-q^n}-\tau(x)\right|^{\nu}\mathrm{d}t\\ &\leq M\frac{[n]_q}{e_{\mu,q}([n]_q\tau(x))}\sum_{k=0}^{\infty}\left(\frac{([n]_q\tau(x))^k}{\gamma_{\mu,q}(k)}\right)^{\frac{2-\nu}{2}}\\ &\quad\times\left(\frac{([n]_q\tau(x))^k}{\gamma_{\mu,q}(k)}\right)^{\frac{\nu}{2}}\left|\frac{1-q^{2\mu\theta_k+k}}{1-q^n}-\tau(x)\right|^{\nu}\mathrm{d}t\\ &\leq M\left(\frac{n}{e_{\mu,q}([n]_q\tau(x))}\sum_{k=0}^{\infty}\frac{([n]_q\tau(x))^k}{\gamma_{\mu,q}(k)}\mathrm{d}t\right)^{\frac{2-\nu}{2}}\\ &\quad\times\left(\frac{[n]_q}{e_{\mu,q}([n]_q\tau(x))}\sum_{k=0}^{\infty}\frac{([n]_q\tau(x))^k}{\gamma_{\mu,q}(k)}\left|\frac{1-q^{2\mu\theta_k+k}}{1-q^n}-\tau(x)\right|^{2}\mathrm{d}t\right)^{\frac{\nu}{2}}\\ &=M\left(L^{\tau}_{n,q}(\tau(t)-\tau(x))^2;x\right)^{\frac{\nu}{2}}\end{aligned}$$

which completes the proof. □

Let

$$C_B^2(\mathbb{R}^+) = \{g \in C_B(\mathbb{R}^+) : g', g'' \in C_B(\mathbb{R}^+)\} \tag{4.9}$$

with the norm

$$\| g \|_{C_B^2(\mathbb{R}^+)} = \| g \|_{C_B(\mathbb{R}^+)} + \| g' \|_{C_B(\mathbb{R}^+)} + \| g'' \|_{C_B(\mathbb{R}^+)}, \tag{4.10}$$

also

$$\| g \|_{C_B(\mathbb{R}^+)} = \sup_{x \in \mathbb{R}^+} | g | . \tag{4.11}$$

**Theorem 4.3** *Let $L_{n,q}^{\tau}(. \, ; \, .)$ be the operator defined in* (2.1). *Then, for any $g \in C_B^2(\mathbb{R}^+)$, we have*

$$| L_{n,q}^{\tau}(g; x) - g | \leq \frac{\delta_{n,\tau(x)}}{2} \| g \|_{C_B^2(\mathbb{R}^+)},$$

*where $\delta_{n,\tau(x)}$ is given in Theorem 4.1.*

***Proof*** Let $g \in C_B^2(\mathbb{R}^+)$. Then by using the generalized mean value theorem in the Taylor series expansion, we have

$$g(\tau(t)) = g + (\tau(t) - \tau(x)).g' + \frac{(\tau(t) - \tau(x))^2}{2}.g''(\psi), \quad \psi \in (\tau(t), \tau(x)).$$

By applying linearity property on $L_{n,q}^{\tau}$, we have

$$L_{n,q}^{\tau}(g, x) - g = g'.L_{n,q}^{\tau}((\tau(t) - \tau(x)); x) + \frac{g''(\psi)}{2}.L_{n,q}^{\tau}\left((\tau(t) - \tau(x))^2; x\right),$$

which imply that

$$| L_{n,q}^{\tau}(g; x) - g | \leq \left([1 + 2\mu]_q \frac{\tau(x)}{[n]_q}\right) \frac{\| g'' \|_{C_B(\mathbb{R}^+)}}{2}.$$

From (4.10), we have

$$\| g'' \|_{C_B[0,\infty)} \leq \| g \|_{C_B^2[0,\infty)} .$$

Thus

$$| L_{n,q}^{\tau}(g; x) - g | \leq \left([1 + 2\mu]_q \frac{\tau(x)}{[n]_q}\right) \frac{\| g \|_{C_B^2(\mathbb{R}^+)}}{2}.$$

This completes the proof from (2) of Lemma 2.2. □

Peetre's $K$-functional is defined by

$$K_2(f,\delta)=\inf_{C_B^2(\mathbb{R}^+)}\left\{\left(\| f-g \|_{C_B(\mathbb{R}^+)} +\delta \| g'' \|_{C_B^2(\mathbb{R}^+)}\right) : g\in \mathcal{W}^2\right\}, \tag{4.12}$$

where

$$\mathcal{W}^2=\left\{g\in C_B(\mathbb{R}^+) : g', g''\in C_B(\mathbb{R}^+)\right\}. \tag{4.13}$$

There exists a positive constant $C>0$ such that

$$K_2(f,\delta)\le C\omega_2(f,\delta^{\frac{1}{2}}),\quad \delta>0,$$

where the second-order modulus of continuity is given by

$$\omega_2(f,\delta^{\frac{1}{2}})=\sup_{0<h<\delta^{\frac{1}{2}}}\ \sup_{x\in\mathbb{R}^+} \mid f(x+2h)-2f(x+h)+f(x)\mid . \tag{4.14}$$

**Theorem 4.4** *Let $L_{n,q}^{\tau}(.\ ;\ \ .)$ be the operator defined in* (2.1). *Then, for any continuously differentiable function $\tau(x)$ on $\mathbb{R}^+$ and $f\in C_B(\mathbb{R}^+)$, we have*

$$\mid L_{n,q}^{\tau}(f;x)-f\mid \le 2M\left\{\omega_2\left(f;\sqrt{\frac{\delta_{n,\tau(x)}}{4}}\right)+\min\left(1,\frac{\delta_{n,\tau(x)}}{4}\right)\| f \|_{C_B(\mathbb{R}^+)}\right\},$$

*where $M$ is a positive constant, $\delta_{n,\tau(x)}$ is given in Theorem 4.3, and $\omega_2(f;\delta)$ is the second-order modulus of continuity of the function $f$ defined in* (4.14).

***Proof*** We prove this by using Theorem 4.3. Let us write

$$\begin{aligned}\mid L_{n,q}^{\tau}(f;x)-f\mid &\le \mid L_{n,q}^{\tau}(f-g;x)\mid + \mid L_{n,q}^{\tau}(g;x)-g\mid + \mid f-g\mid\\ &\le 2\| f-g \|_{C_B(\mathbb{R}^+)} +\frac{\delta_{n,\tau(x)}}{2}\| g \|_{C_B^2(\mathbb{R}^+)} .\end{aligned}$$

From (4.10), we clearly have

$$\| g \|_{C_B[0,\infty)}\le\| g \|_{C_B^2[0,\infty)} .$$

Therefore,

$$\mid L_{n,q}^{\tau}(f;x)-f\mid \le 2\left(\| f-g \|_{C_B(\mathbb{R}^+)} +\frac{\delta_{n,\tau(x)}}{4}\| g \|_{C_B^2(\mathbb{R}^+)}\right),$$

where $\delta_{n,\tau(x)}$ is given in Theorem 4.1.

By taking infimum over $g\in C_B^2(\mathbb{R}^+)$ and using (4.12), we get

$$\mid L_{n,q}^{\tau}(f;x)-f\mid \le 2K_2\left(f;\frac{\delta_{n,\tau(x)}}{4}\right).$$

Now for an absolute constant $Q > 0$ in [10] we use the relation

$$K_2(f;\delta) \leq Q\{\omega_2(f;\sqrt{\delta}) + \min(1,\delta) \parallel f \parallel\}.$$

This completes the proof. □

## 5 Concluding Remark

In our investigation, we constructed the $q$-analogue of Szász operators generated by Dunkl exponential function based on a continuously differentiable function $\tau$ on $[0,\infty)$ such that $\tau(0) = 0$ and $\inf_{x\in\mathbb{R}^+} \tau'(x) \geq 1$ and then obtained auxiliary lemmas. We also obtained the uniform and weighted approximation theorems of our aforesaid operators $L^{\tau}_{n,q}$ defined in (2.1). Further, we established the rate of convergence of $L^{\tau}_{n,q}$ by means of modulus of continuity and Lipschitz-type functions.

However, if we take $\tau = t$ then the operators $L^{\tau}_{n,q}$ coincide with the operators $D_{n,q}$ defined in (1.7) (see [15]) which means that our operators are non-trivial extension of $D_{n,q}$ and hence our results.

## References

1. A. Alotaibi, M. Mursaleen, Approximation of Jakimovski-Leviatan-Beta type integral operators via $q$-calculus. AIMS Math. **5**(4), 3019–3034 (2020)
2. A. Alotaibi, On the approximation by bivariate Szász-Jakimovski-Leviatan-type operators of unbounded sequences of positive numbers. Mathematics **11**, Article 1009 (2023)
3. A. Aral, O. Doğru, Bleimann Butzer and Hahn operators based on $q$-integers. J. Inequal. Appl. **2007**, Article ID 79410 (2007)
4. A. Aral, D. Inoan, I. Raşa, On the generalized Szász-Mirakyan operators. Results. Math. **65**, 441–452 (2014)
5. R. Aslan, On a Stancu form Szász-Mirakjan-Kantorovich operators based on shape parameter $\lambda$. Adv. Stud.: Euro-Tbilisi Math. J. **15**(1), 151–166 (2022)
6. R. Aslan, Rate of approximation of blending type modified univariate and bivariate $\lambda$-Schurer-Kantorovich operators. Kuwait J. Sci. **51**(1), 100168 (2024)
7. S.N. Bernstein, Démonstration du théoréme de Weierstrass fondée sur le calcul des probabilités. Commun. Soc. Math. Kharkow **2**(13), 1–2 (2012)
8. D. Cárdenas-Morales, P. Garrancho, I. Raşa, Bernstein-type operators which preserve polynomials. Comput. Math. Appl. **62**, 158–163 (2011)
9. B. Cheikh, Y. Gaied, M. Zaghouani, A $q$-Dunkl-classical $q$-Hermite type polynomials. Georgian Math. J. **21**(2), 125–137 (2014)
10. A. Ciupa, A class of integral Favard-Szász type operators. Stud. Univ. Babeş-Bolyai, Math. **40**(1), 39–47 (1995)
11. O. Duman, M.A. Özarslan, Szász-Mirakjan type operators providing a better error estimation. Appl. Math. Lett. **20**, 1184–1188 (2007)
12. O. Duman, M.A. Özarslan, H. Aktuğlu, Better error estimation for Szász-Mirakjan-Beta operators. J. Comput. Anal. Appl. **10**, 53–59 (2008)

13. A.D. Gadžiev, The convergence problem for a sequence of positive linear operators on unbounded sets and theorems analogues to that of P. P. Korovkin, Dokl. Akad. Nauk SSSR, 218, 1001–1004. Also in Soviet Math Dokl. **15**(1974), 1433–1436 (1974). ((in English))
14. A.D. Gadžiev, Theorems of the type of P.P. Korovkin's theorems (in Russian). Math. Z. **205**, 781–786 (1976); translated in Math. Notes **20** (5–6), 995–998 (1977)
15. G. İçöz, B. Çekim, Dunkl generalization of Szász operators via $q$-calculus. J. Inequal. Appl. **2015**, Article 284 (2015)
16. V. Kac, P. Cheung, *Quantum Calculus* (Springer, New York, 2002)
17. P.P. Korovkin, *Linear Operators and Approximation Theory* (Hindustan, Delhi, 1960)
18. B. Lenze, On Lipschitz-type maximal functions and their smoothness spaces. Indag. Math. **91**, 53–63 (1988)
19. A. Lupaş, A $q$-analogue of the Bernstein operator, in *Seminar on Numerical and Statistical Calculus*, vol. 9 (University of Cluj-Napoca, Cluj-Napoca, 1987), pp. 85–92
20. S.A. Mohiuddine, N. Ahmad, F. Özger, A. Alotaibi, B. Hazarika, Approximation by the parametric generalization of Baskakov-Kantorovich operators linking with Stancu operators. Iran. J. Sci. Technol. Trans. Sci. **45**, 593–605 (2021)
21. S.A. Mohiuddine, F. Özger, Approximation of functions by Stancu variant of Bernstein–Kantorovich operators based on shape parameter $\alpha$. Rev. Real Acad. Cienc. Exactas Fis. Nat. Ser. A-Mat. **114**, 70 (2020)
22. M. Mursaleen, K.J. Ansari, A. Khan, Approximation properties and error estimation of $q$-Bernstein shifted operators. Numer. Algorithms **84**, 207–227 (2020)
23. M. Mursaleen, K.J. Ansari, M. Nasiruzzaman, Approximation by $q$-analogue of Jakimovski-Leviatan operators involving $q$-Appell polynomials. Iranian J. Sci. Tech.-A (Sci.). **41**, 891–900 (2017)
24. M.A. Mursaleen, M. Heshamuddin, N. Rao, B.K. Sinha, A.K. Yadav, Hermite polynomials linking Szász-Durrmeyer operators. Comput. Appl. Math. **43**(4), 223 (2024)
25. M. Mursaleen, T. Khan, M. Nasiruzzaman, Approximating properties of generalized Dunkl analogue of Szász operators. Appl. Math. Inf. Sci. **10**(6), 1–8 (2016)
26. M. Mursaleen, M. Nasiruzzaman, H.M. Srivastava, Approximation by bicomplex beta operators in compact BC-disks. Math. Meth. Appl. Sci. **39**, 2916–2929 (2016)
27. M. Nasiruzzaman, A.F. Aljohani, Approximation by parametric extension of Szász-Mirakjan-Kantorovich operators Involving the Appell polynomials. J. Funct. Spaces **2020**, Article ID 8863664 (2020)
28. M. Nasiruzzaman, Approximation properties by Szász operators to bivariate functions via Dunkl analogue. Iran. J. Sci. Technol. Trans. Sci. **45**, 259–269 (2021)
29. M. Nasiruzzaman, M. Mursaleen, Approximation by Jakimovski-Leviatan-Beta operators in weighted space. Adv. Diff. Equ. **2020**, Article 393 (2020)
30. F. Özger, H.M. Srivastava, S.A. Mohiuddine, Approximation of functions by a new class of generalized Bernstein-Schurer operators. Rev. Real Acad. Cienc. Exactas Fis. Nat. Ser. A-Mat. 114:173 (2020)
31. G.M. Phillips, Bernstein polynomials based on the $q$-integers. Ann. Numer. Math. **4**, 511–518 (1997)
32. M. Rosenblum, Generalized Hermite polynomials and the Bose-like oscillator calculus. Oper. Theory Adv. Appl. **73**, 369–396 (1994)
33. D. Söylemez, On $q$-Bleimann, Butzer, and Hahn-type operators. Abstr. Appl. Anal. **2015**, Article ID 480925, 7 pages (2015)
34. L.T. Su, R. Aslan, F.S. Zheng, M. Mursaleen, On the Durrmeyer variant of $q$-Bernstein operators based on the shape parameter $\lambda$. J. Inequal. Appl. **2023**, Article 56 (2023)
35. S. Sucu, Dunkl analogue of Szász operators. Appl. Math. Comput. **244**, 42–48 (2014)
36. O. Szász, Generalization of S. Bernstein's polynomials to the infinite interval. J. Res. Natl. Bur. Stand. **45**, 239–245 (1950)
37. A. Wafi, N. Rao, A generalization of Szász-type operators which preserves constant and quadratic test functions. Cogent Math. **3**, Article ID 1227023 (2016)

# A Study on Range and Kernel of Toeplitz and Hankel Operators on the Bergman Space

**Chinmayee Padhy, Pabitra Kumar Jena, and S. K. Paikray**

**Abstract** In this article, we establish the inclusion relationship between the kernel and range of little Hankel operators and Toeplitz operators on the Bergman space. Further, we expound on the conditions for finite rank operators and invertibility of operators on $A^2(\mathbb{D})$. Moreover, we explore some necessary and sufficient conditions under which the bounded linear operators on the Bergman space $A^2(\mathbb{D})$ are positive.

**Keywords** Operator ranges · Toeplitz operators · Little Hankel operators · Bergman space

**2010 Mathematics Subject Classification** 47B38 · 47B35 · 47B33

## 1 Introduction

Let $dA(z) = \frac{1}{\pi}dxdy$ be the normalized Lebesgue area measure on the open unit disk $\mathbb{D}$, whose measure of the disk is equal to 1. The Bergman space $A^2(\mathbb{D})$, which is the closed subspace of $L^2(\mathbb{D}, dA)$ consisting of all holomorphic functions on $\mathbb{D}$.

The reproducing kernel $K_z$, for $z \in \mathbb{D}$ of Bergman space $A^2(\mathbb{D})$ is a function which satisfies $f(z) = \langle f, K_z\rangle \ \forall f \in A^2(\mathbb{D})$ and $k_z = \frac{K_z}{\|K_z\|_2}$ denotes the normalized reproducing kernel. For any integer $n \geq 0$, let $e_n(z) = \sqrt{n+1}z^n$. Then, $\{e_n\}_{n=0}^{\infty}$ forms an orthonormal basis for $A^2(\mathbb{D})$. Given $\phi \in L^\infty(\mathbb{D})$, the Toeplitz operator $T_\phi$ is defined on $A^2(\mathbb{D})$ by

$$T_\phi h = P(\phi h).$$

C. Padhy
Department of Basic Science and Humanities, Silicon Institute of Technology, Sason, Sambalpur, Odisha, India

P. K. Jena
P. G. Department of Mathematics, Berhampur University, Berhampur, Ganjam, Odisha, India

S. K. Paikray (✉)
Department of Mathematics, Veer Surendra Sai University of Technology, Burla, Odisha, India
e-mail: skpaikray_math@vssut.ac.in



S. A. Mohiuddine et al. (eds.), *Positive Operators and Fixed-Point Theorems with Applications*, Industrial and Applied Mathematics,
https://doi.org/10.1007/978-981-96-2476-8_8

Thus, we have

$$(T_\phi h)(w) = \int_{\mathbb{D}} \frac{\phi(z)h(z)}{(1-\overline{z}w)^2} dA(z),$$

for $h \in A^2(\mathbb{D})$ and $w \in \mathbb{D}$.

Similarly, one can define little Hankel operators on $A^2(\mathbb{D})$. For $\phi \in L^\infty(\mathbb{D})$, the little Hankel operator $S_\phi$ on $A^2(\mathbb{D})$ with symbol $\phi$ is the operator defined by $S_\phi f = PJ(\phi f)$, where $J : L^2(\mathbb{D}, dA) \to L^2(\mathbb{D}, dA)$ is defined as $Jf(z) = f(\overline{z})$. To know more regarding Bergman space see [12].

The algebra of bounded, linear operators from a Hilbert space $H$ into itself is denoted by $\mathcal{L}(H)$. An operator $A$ is called positive and denoted by $A \geq 0$ if $\langle Ax, x\rangle \geq 0$ for all $x \in H$ and $\sigma(A)$ denotes the spectrum of $A$. We define the Berezin transform of a bounded linear operator Q on $A^2(\mathbb{D})$ to be the function $\widetilde{Q}$ defined on $\mathbb{D}$ by

$$\widetilde{Q}(w) = \langle Qk_w, k_w\rangle, \text{ for } w \in \mathbb{D}.$$

Let $\widetilde{\phi}(w) = \langle T_\phi k_w, k_w\rangle$ for $w \in \mathbb{D}$. That is, $\widetilde{\phi} = \widetilde{T_\phi}$. The Schatten $p$ class operator is denoted by $S_p(H)$ of Hilbert space $H$ for $0 < p < \infty$, is the space of all compact operators $A$ on $H$. Additionally, a few of our main results have some advanced applications which are connected to the recent findings of Padhy et al. [9–11], Jung et al. [7], and Gurdal [4].

In this article, we prove the inclusion relationship between the kernel and range of little Hankel and Toeplitz operators on Bergman space. The kernel of little Hankel operators is studied in Sect. 2. Section 3 deals with finite rank and invertibility of operators on Bergman space. We established some results to show the positivity of operators on Bergman Space in Sect. 4.

## 2 Kernel of Little Hankel Operators

In this section, several inclusion relationships between kernel and range of little Hankel operators are established. For a little Hankel operator $S_\phi \in A^2(\mathbb{D})$, $S_\phi^* = S_{\phi^+}$, where $\phi^+(z) = \overline{\phi(\overline{z})}$ for every $\phi \in L^\infty(\mathbb{D})$.

An operator $Q$ is said to be asymptotic regular iff $\lim_{n\to\infty} \|Q^n g - Q^{n+1}g\| = 0$ for $g \in A^2(\mathbb{D})$.

**Lemma 2.1** *Let $\psi \in L^\infty(\mathbb{D})$ and $S_{\frac{1+\psi}{2}}$ be asymptotic regular. Then Kernel $(I - S_\psi) \cap$ Range $(I - S_\psi) = \{0\}$.*

***Proof*** Let $W = S_{\frac{1+\psi}{2}}$. Then $2(I - W) = I - S_\psi$. Therefore the *Kernel* $(I - W) =$ *Kernel* $(I - S_\psi)$ and *Range* $(I - W) =$ *Range* $(I - S_\psi)$. Assume that $g \in$ *Kernel* $(I - W) \cap$ *Range* $(I - W)$. Therefore $(I - W)g = 0$, so $Wg = g$. Again $W^2 g = W(Wg) = Wg = g$. By using mathematical induction we get $W^n g = g$, for every $n \in \mathbb{N}$. Since $g \in$ *Range* $(I - W)$, so $(I - W)h = g$ for some $h \in A^2(\mathbb{D})$. Thus $g = h - Wh$. Hence $g = W^n h - W^{n+1} h$ that implies $g = S^n_{\frac{1+\psi}{2}} h - S^{n+1}_{\frac{1+\psi}{2}} h$. As $S_{\frac{1+\psi}{2}}$ is asymptotic regular, $\|S^n_{\frac{1+\psi}{2}} h - S^{n+1}_{\frac{1+\psi}{2}} h\| \to 0$ as $n \to \infty$, which implies that $g = 0$. So *Kernel* $(I - W) \cap$ *Range* $(I - W) = \{0\}$. Thus *Kernel* $(I - S_\psi) \cap$ *Range* $(I - S_\psi) = \{0\}$. □

**Theorem 2.1** *Let $\psi \in L^\infty(\mathbb{D})$ and $S_{\frac{1+\psi}{2}}$ be asymptotic regular. Then Kernel $(I - S_{\psi^+}) =$ Kernel $(I - S_{\psi^+})^2$.*

***Proof*** By Lemma 2.1, we have *Kernel* $(I - S_{\psi^+}) \cap$ *Range* $(I - S_{\psi^+}) = \{0\}$. It is clear that *Kernel* $(I - S_{\psi^+}) \subseteq$ *Kernel* $(I - S_{\psi^+})^2$. Now we have only to prove the other inclusion, that is, *Kernel* $(I - S_{\psi^+})^2 \subseteq$ *Kernel* $(I - S_{\psi^+})$. Let $g \in$ *Kernel* $(I - S_{\psi^+})^2$. Then for some $h \in A^2(\mathbb{D})$, $h = (I - S_{\psi^+})g$. So $h \in$ *Kernel* $(I - S_{\psi^+}) \cap$ *Range* $(I - S_{\psi^+})$, therefore, $h = 0$. Hence, $g \in$ *Kernel* $(I - S_{\psi^+})$ which implies that *Kernel* $(I - S_{\psi^+})^2 \subseteq$ *Kernel* $(I - S_{\psi^+})$. Hence the theorem is proved.□

For $Q_1, Q_2 \in \mathcal{L}(H)$, we use the notation $[Q_1, Q_2] = Q_1 Q_2 - Q_2 Q_1$ in the remaining part of the paper.

**Theorem 2.2** *Let $\psi \in L^\infty(\mathbb{D})$ and let $S_\psi = VR$ be the polar decomposition of $S_\psi$ with $V$ as partial isometry and $R$ as positive, where Kernel $V =$ Kernel $R$. Then $[S_{\psi^+} S_\psi, S_\psi S_{\psi^+}] = 0$ if and only if $[R, VRV^*] = 0$.*

***Proof*** We first recall that $V^*V$ is the Bergman projection onto the Range R. Therefore $R^*VR = R$ since $R \geq 0$. Hence $[S_{\psi^+} S_\psi, S_\psi S_{\psi^+}] = 0 \Leftrightarrow [R^2, VR^2V^*] = 0 \Leftrightarrow [R, (VR^2V^*)^{\frac{1}{2}}] = 0 \Leftrightarrow [R, VRV^*] = 0$. □

**Theorem 2.3** *Let $\psi \in L^\infty(\mathbb{D})$ and $V$ and $Q$ be the operators in $\mathcal{L}(A^2(\mathbb{D}))$ such that $S^2_\psi = (VQ)^2$ and $S^3_\psi = (VQ)^3$. If Kernel $S_\psi \subset$ Kernel $S_{\psi^+}$ and Kernel $VQ \subset$ Kernel $(VQ)^*$ then $S_\psi = VQ$.*

***Proof*** Let $R = S_\psi - VQ$. Now $S^2_\psi R = S^3_\psi - S^2_\psi(VQ) = S^3_\psi - (VQ)^3 = 0$. As *Kernel* $S_\psi \subset$ *Kernel* $S_{\psi^+}$ that implies $S_{\psi^+} S_\psi R = 0$. Therefore, $(S_\psi R)^* S_\psi R = 0$ and $S_\psi R = 0$. Hence $S^*_\psi R = 0$. Similarly, $(VQ)^2 R = (VQ)^2 S_\psi - (VQ)^3 = 0$. As *Kernel* $VQ \subset$ *Kernel* $(VQ)^*$ that implies $(VQ)^*(VQ)R = 0$. Thus, $(VQ)R = 0$ and $(VQ)^* R = 0$. Therefore, $R^* R = (S_\psi - VQ)^* R = S_{\psi^+} R - (VQ)^* R = 0$. That implies $\|R\|^2 = 0$, which implies $R = 0$. Hence $S_\psi = VQ$. □

## 3 On Operator Ranges

In this section, we obtained the conditions for finite rank and invertibility of an operator on the Bergman space.

**Proposition 3.1** *Let $\phi$ , $\psi \in L^\infty(\mathbb{D})$. If $T_\phi$ and $S_\psi$ be isometries on $A^2(\mathbb{D})$. Then the following assertions are pairwise equivalent:*

(i) $\textit{Range } T_\phi = \textit{Kernel } S_\psi^*$.
(ii) $\textit{Range } S_\psi = \textit{Kernel } T_\phi^*$.
(iii) $T_\phi T_\phi^* + S_\psi S_\psi^* = I$.

***Proof*** The equivalence between **(i)** and **(ii)** can be easily verified by recalling that $\textit{Kernel } T_\phi = (\textit{Range } T_\phi^*)^\perp$ and $\overline{\textit{Range } T_\phi} = (\textit{Kernel } T_\phi^*)^\perp$ and the fact that isometries have a closed range.

Suppose any of the equivalent assertions **(i)** and **(ii)** hold and take an arbitrary $f = g + h \in A^2(\mathbb{D}) = \textit{Range } T_\phi + \textit{Range } T_\phi^\perp = \textit{Kernel } S_\psi^* + \textit{Range } S_\psi$ so that $g \in \textit{Range } T_\phi = \textit{Kernel } S_\psi^*$ and $h \in \textit{Range } T_\phi^\perp = \textit{Range } S_\psi = \textit{Kernel } T_\phi^*$. Thus

$$\begin{aligned}(T_\phi T_\phi^* + S_\psi S_\psi^*)f &= T_\phi T_\phi^* g + T_\phi T_\phi^* h + S_\psi S_\psi^* g + S_\psi S_\psi^* h \\ &= T_\phi T_\phi^* g + S_\psi S_\psi^* h \\ &= T_\phi T_\phi^* T_\phi f_1 + S_\psi S_\psi^* S_\psi f_2 \\ &\quad (\textit{for some } f_1 \textit{ and } f_2 \in A^2(\mathbb{D}) \textit{ such that } g = T_\phi f_1 \textit{ and } h = S_\psi f_2) \\ &= T_\phi f_1 + S_\psi f_2 \\ &= g + h = f\end{aligned}$$

that is assertion **(iii)** holds. Conversely, suppose (iii) holds. Let $f \in \textit{Range } T_\phi$, so that $T_\phi g = f$ for some $g \in A^2(\mathbb{D})$. Then, $T_\phi T_\phi^* f = T_\phi T_\phi^* T_\phi g = T_\phi g = f$ and hence $f = T_\phi T_\phi^* f + S_\psi S_\psi^* f = f + S_\psi S_\psi^* f$ that implies $S_\psi S_\psi^* f = 0$, that is, $f \in \textit{Kernel } S_\psi S_\psi^* = \textit{Kernel } S_\psi^*$. Hence, $\textit{Range } T_\phi \subseteq \textit{Kernel } S_\psi^*$. On the other hand, let $h \in \textit{Kernel } S_\psi^*$. Then $h = (T_\phi T_\phi^* + S_\psi S_\psi^*)h = T_\phi T_\phi^* h \in \textit{Range } T_\phi$ and so $\textit{Kernel } S_\psi^* \subseteq \textit{Range } T_\phi$. Hence **(iii)** implies **(i)**. □

**Theorem 3.1** *Let $\phi \in L^\infty(\mathbb{D})$ with $\|\phi\|_\infty \le 1$ and $V$ be an isometry. Then $T_\phi - V$ is a finite rank operator if and only if $T_\phi = XV$ and $X - I$ is a finite rank operator, where $X$ is a contraction operator in $\mathcal{L}(A^2(\mathbb{D}))$.*

***Proof*** Let $Y = T_\phi - V$ and put $W = \textit{Range } Y$ (which has finite dimension). It can be observed that $W \cap \textit{Kernel } V^* = \{0\}$. But, if $f = Yg$ is in $W \cap \textit{Kernel } V^*$, then

$$\begin{aligned}\|f\|^2 &= \langle V^*Yg, g\rangle + \langle g, V^*Yg\rangle + \|Yg\|^2 \\ &= \|(V+Y)g\|^2 - \|Vg\|^2 \le 0.\end{aligned}$$

That implies $f = 0$.

Let $\check{P}$ be the orthogonal projection onto $W$. Since $Y^*(I - \check{P}) = (I - \check{P})Y = 0$, the inequality

$$(V+Y)^*(V+Y)^* \le I = VV^*$$

is equivalent to

$$(V^* + Y^*)\check{P}(V+Y) + V^*(I - \check{P})V \le V^*\check{P}V + V^*(I - \check{P})V$$

or

$$(V^* + Y^*)\breve{P}(V + Y) \le V^*\breve{P}V.$$

Thus the map $X_0 : \breve{P}Vf \mapsto \breve{P}(V + Y)f$ is a contraction defined on $\breve{P}V(A^2(\mathbb{D}))$. But $\breve{P}V(A^2(\mathbb{D})) = W$ since $W \cap \mathit{Kernel}\ V^* = \{0\}$. Then $X \in \mathcal{L}(A^2(\mathbb{D}))$ defined by $X = I - \breve{P} + X_0\breve{P}$ is the required contraction. The converse is trivial. □

If $Q^*(Q^*Q - QQ^*)Q \ge 0$, that is, $\|Q^*Qf\| \le \|QQf\|$ for all $f \in H$, then the operator $Q \in \mathcal{L}(\mathcal{H})$ is called quasi-hyponormal.

**Lemma 3.1** *Let $\phi \in L^\infty(\mathbb{D})$ and $T_\phi \in \mathcal{L}(A^2(\mathbb{D}))$ be a quasi-hyponormal operator. Let $\{g_n\}_{n=0}^\infty$ be a sequence in $A^2(\mathbb{D})$ such that $Tg_{n+1} = g_n$ for all $n \ge 0$. Then either $\|g_0\| \ge \|g_1\| \ge \|g_3\| \cdots$ or $\|g_n\| \to \infty$ as $n \to \infty$.*

***Proof*** For any $g \in A^2(\mathbb{D})$,

$$\begin{aligned}\|T_\phi g\| &= \langle T_\phi g, T_\phi g\rangle^{\frac{1}{2}} = \langle T_{\overline{\phi}}T_\phi g, g\rangle^{\frac{1}{2}} \le (\|T_{\overline{\phi}}T_\phi g\|\|T_\phi\|)^{\frac{1}{2}}\\ &\le (\|T_\phi^2 g\|\|g\|)^{\frac{1}{2}} \le \frac{1}{2}(\|T_\phi^2 g\| + \|g\|).\end{aligned}$$

Letting $g = g_{n+2}$, we see that $\|g_{n+1}\| \le \frac{1}{2}(\|g_n\| + \|g_{n+2}\|)$, so the sequence $\{\|g_n\|\}$ is convex and the conclusion follows from [13]. □

**Lemma 3.2** *Suppose $S \in \mathcal{L}(A^2(\mathbb{D}))$ is invertible, and $T \in \mathcal{L}(A^2(\mathbb{D}))$ is a quasi-hyponormal, and $S_\phi \in \mathcal{L}(A^2(\mathbb{D}))$ satisfies $TS_\phi = S_\phi S$. Then*

$$\|S_\phi S^{-1}h\| \le \|S^{-1}\|\|S_\phi h\| \ \textit{for all}\ h \in A^2(\mathbb{D}).$$

***Proof*** Let $c = \|S^{-1}\| > 0$. Fix $h \in A^2(\mathbb{D})$ and define $g_n = c^{-n}S_\phi S^{-n}h$ for $n \ge 0$. Then $cTg_{n+1} = g_n$ and $\|g_n\| \le \|S_\phi\|\|h\|$ for all $n \ge 0$. Since $cT$ is quasi-hyponormal, $\|g_0\| \ge \|g_1\| \ge \|g_2\| \cdots$ by Lemma 3.1, and the first inequality in this chain, $\|g_1\| \le \|g_0\|$, shows that $\|S_\phi S^{-1}h\| \le \|S^{-1}\|\|S_\phi h\|$. □

**Proposition 3.2** *If $S \in \mathcal{L}(A^2(\mathbb{D}))$ is invertible, $T \in \mathcal{L}(A^2(\mathbb{D}))$ is a quasi-hyponormal, $S_\phi \in \mathcal{L}(A^2(\mathbb{D}))$ has dense range and satisfies $TS_\phi = S_\phi S$, then $T$ is invertible.*

***Proof*** Since $TS_\phi = S_\phi S$, the range of $T$ contains a range of $S_\phi$, so $T$ has a dense range. It remains only to show that $T$ is bounded below, and by continuity, it will suffice to show that $T$ is bounded below on the range of $T$. By Lemma 3.2,

$$\|S_\phi S^{-1}h\| \le \|S^{-1}\|\|S_\phi h\| = \|S^{-1}\|\|S_\phi SS^{-1}h\| = \|S^{-1}\|\|TS_\phi S^{-1}h\|$$

for all $h \in A^2(\mathbb{D})$. Put $h_1 = S^{-1}h$. Then $\|S^{-1}\|^{-1}\|S_\phi h_1\| \le \|TS_\phi h_1\|$, and so $T$ is bounded below on the range of $S_\phi$. □

## 4 Positive Operators

In this section, we shall establish the necessary and sufficient conditions under which Toeplitz and little Hankel operators on $A^2(\mathbb{D})$ are positive.

**Theorem 4.1** *Let $A, B \in \mathcal{L}(A^2(\mathbb{D}))$ and $\phi \in L^\infty(\mathbb{D})$ such that Kernel $T_{\overline{\phi}} = \{0\}$. Then $A \geq B$ if and only if $T_{\overline{\phi}} A T_\phi \geq T_{\overline{\phi}} B T_\phi$.*

***Proof*** Let $A \geq B$. Then we get the following relation:

$$\langle T_{\overline{\phi}} A T_\phi k_z, k_z\rangle = \langle A T_\phi k_z, T_\phi k_z\rangle.$$

For $g = T_\phi k_z$ it follows that $\langle A T_\phi k_z, T_\phi k_z\rangle = \langle Ag, g\rangle \geq \langle Bg, g\rangle = \langle B T_\phi k_z, T_\phi k_z\rangle = \langle T_{\overline{\phi}} B T_\phi k_z, k_z\rangle$.

Conversely, let $T_{\overline{\phi}} A T_\phi \geq T_{\overline{\phi}} B T_\phi$. Then we get the following relation: $\langle T_{\overline{\phi}} A T_\phi k_z, k_z\rangle \geq \langle T_{\overline{\phi}} B T_\phi k_, k_z\rangle$. That implies $\langle A T_\phi k_z, T_\phi k_z\rangle \geq \langle B T_\phi k_z, T_\phi k_z\rangle$. Since $T_\phi$ has dense range in $A^2(\mathbb{D})$, for $g = T_\phi k_z$, we have $\langle Ag, g\rangle \geq \langle Bg, g\rangle$ which implies that $A \geq B$. □

**Proposition 4.1** *Let $A, B \in \mathcal{L}(A^2(\mathbb{D}))$ be such that $A$ and $B$ are positive and for $c_n > 0, c_i \geq 0$*

$$c_1 A + c_2 A^2 + \cdots + c_n A^n \geq c_1 B + c_2 B^2 + \cdots + c_n B^n. \tag{4.1}$$

*If $A = aP$ where $P$ is the Bergman projection and for a scalar $a > 0$, then $A \geq B$.*

***Proof*** We may assume $a = 1$; in fact, when $a \neq 1$ we only need to rewrite Eq. (4.1) as

$$(c_1 a)P + \cdots + (c_n a)P^n \geq (c_1 a)(\frac{B}{a}) + \cdots + c_n a^n (\frac{B}{a})^n$$

to get $A \geq B$. Set $b_i = \frac{c_i}{(c_1 + c_2 + \cdots c_n)}$. Then Eq. (4.1) gives $P \geq b_1 B + b_2 B^2 + \cdots b_n B^n$. By Jensen's inequality, $\langle B^j k_z, k_z\rangle \geq \langle B k_z, k_z\rangle^j$. So we get

$$\langle P k_z, k_z\rangle \geq \sum b_j \langle B k_z, k_z\rangle^j, \sum_{j=1}^{n} b_j = 1$$

from which it follows that if $\langle P k_z, k_z\rangle = 1$ then $\langle B k_z, k_z\rangle \leq 1$ and that if $\langle P k_z, k_z\rangle = 0$, then $\langle B k_z, k_z\rangle = 0$. Therefore, $PBP \leq P$ and $(I - P)B(I - P) = 0$ which implies $B(I - P) = 0$. Thus, we obtain $A = P \geq PBP = B$. □

**Proposition 4.2** *Let $A, B \in \mathcal{L}(A^2(\mathbb{D}))$ be such that $A$ is self-adjoint, $B$ is positive, and $B^2 \geq A^2$. Then $B \geq A$.*

***Proof*** Choose $\lambda < 0$. Then $\|(B-\lambda)k_z\|^2 - \|Ak_z\|^2 = \langle (B^2-A^2)k_z, k_z\rangle - 2\lambda\langle Bk_z, k_z\rangle + \lambda^2 \geq \lambda^2$. Hence $\|(B-A-\lambda)k_z\| \geq \|(B-\lambda)k_z\| - \|Ak_z\| \geq \frac{\lambda^2}{\|B-\lambda\|+\|A\|} > 0$. Since $B-A$ is Hermitian, we have $\sigma(B-A) \geq 0$. Consequently, $B \geq A$. □

**Theorem 4.2** *Let $\phi \in L^\infty(\mathbb{D})$. Then $T_\phi$ is positive-normal if and only if $T_\phi T_\phi^* \leq T_\phi^* P T_\phi$, where $P$ is a positive operator belongs to $\mathcal{L}(A^2(\mathbb{D}))$.*

***Proof*** It is sufficient to show that if there exists a positive operator $P \in \mathcal{L}(A^2(\mathbb{D}))$ such that $T_\phi T_\phi^* \leq T_\phi^* P T_\phi$, then $T_\phi$ is positive-normal. For any $f \in A^2(\mathbb{D})$,

$$\begin{aligned}
\|T_\phi^* f\|^2 &= \langle T_\phi T_\phi^* f, f\rangle \\
&\leq \langle T_\phi^* P T_\phi f, f\rangle \\
&= \langle \sqrt{P} T_\phi f, \sqrt{P} T_\phi f\rangle \\
&= \|\sqrt{P} T_\phi f\|^2 \\
&\leq \|\sqrt{P}\|^2 \|T_\phi f\|^2.
\end{aligned}$$

Hence, for any $f \in A^2(\mathbb{D})$,

$$\|T_\phi^* f\| \leq \|\sqrt{P}\| \|T_\phi f\|.$$

Set $\lambda = \|\sqrt{P}\|$. Then, for any $f \in A^2(\mathbb{D})$,

$$\|T_\phi^* f\| \leq \lambda \|T_\phi f\|.$$

Thus,

$$T_\phi T_\phi^* \leq \lambda^2\, T_\phi^* T_\phi \quad \text{for some} \lambda \geq 0.$$

By [3], there exists $B \in \mathcal{L}(A^2(\mathbb{D}))$ such that $T_\phi = T_\phi^* B$. Therefore,

$$T_\phi T_\phi^* = (T_\phi^* B)(B^* T_\phi) = T_\phi^*(BB^*)T_\phi.$$

So $T_\phi$ is a positive-normal operator with an interrupter $BB^*$. □

**Proposition 4.3** *Let $D = T_\phi T_{\overline{\phi}}$ be a bounded positive operator on $A^2(\mathbb{D})$. Then*

$$\textit{Range}\, D^{\frac{1}{2}} = \textit{Range}\, T_\phi = \{h \in A^2(\mathbb{D}) : \sup\{|\langle h, g\rangle|^2 : g \in A^2(\mathbb{D}), \langle Dg, g\rangle \leq 1\} < \infty\}.$$

***Proof*** Let $h = T_\phi f$, then

$$|\langle h, g\rangle|^2 = |\langle f, T_{\overline{\phi}} g\rangle|^2 \leq \|f\|^2 \langle T_\phi T_{\overline{\phi}} g, g\rangle.$$

On the other hand, if $h$ is in the above given set then $T_{\overline{\phi}} g \mapsto \langle g, h\rangle$ is a bounded linear functional, denote $f$ one of its possible representing vectors. Then $\langle T_{\overline{\phi}} g, f\rangle = \langle g, h\rangle$, that is, $T_\phi f = h$. Since we can replace $T_\phi$ and $T_{\overline{\phi}}$ by $D^{\frac{1}{2}}$ in this argument, the proposition is proved. □

**Proposition 4.4** *If $T_\phi \in \mathcal{L}(A^2(\mathbb{D}))$ is positive-normal, then Kernel $T_\phi$ = Kernel $T_\phi^2$.*

***Proof*** We have to show that *Kernel* $T_\phi^2 \subset$ *Kernel* $T_\phi$. If $f \in$ *Kernel* $T_\phi^2$, then $T_\phi^2 f = 0$. Hence $T_\phi f \in$ *Kernel* $T_\phi$. Since *Kernel* $T_\phi \subset$ *Kernel* $T_\phi^*$, $T_\phi f \in$ *Kernel* $T_\phi^*$. Hence $T_\phi^* T_\phi f = 0$. Now,

$$\|T_\phi f\|^2 = \langle T_\phi f, T_\phi f\rangle = \langle T_\phi^* T_\phi f, f\rangle \leq \|T_\phi^* T_\phi f\| \|f\| = 0.$$

Thus $T_\phi f = 0$ and so we have $f \in$ *Kernel* $T_\phi$. □

**Theorem 4.3** *Let $\phi, \psi \in L^\infty(\mathbb{D})$, where $\phi, \psi$ be nonnegative with $T_\phi \geq c > 0$ and $T_\psi \geq c > 0$. Then*

*(i)* $2c\|T_{\phi-\psi}\| = \|T_\phi^2 - T_\psi^2\|$ *if and only if* $\phi = \psi$.
*(ii)* $2c\|T_{\phi-\psi}\|_p = \|T_\phi^2 - T_\psi^2\|_p < \infty$ *for some* $1 \leq p < \infty$, *if and only if* $\phi = \psi$.

***Proof*** Let $M = T_{\phi-\psi}$. Then $T_\phi^2 - T_\psi^2 = T_\phi M + M T_\psi$. If $2c\|T_{\phi-\psi}\| = \|T_\phi^2 - T_\psi^2\|$, then $2c\|M\| = \|T_\phi M + M T_\psi\|$. We shall show that $M = 0$. Since $M$ is self-adjoint, there is a sequence $\{f_n\}$ of unit vectors in $A^2(\mathbb{D})$ such that $Mf_n - tf_n \to 0$ as $n \to \infty$ where $|t| = \|M\|$. Now

$$\begin{aligned} 2c|t| &= \|T_\phi M + M T_\psi\| \\ &\geq |\langle (T_\phi M + M T_\psi) f_n, f_n\rangle| \\ &= |\langle (M-t)f_n, T_\phi f_n\rangle + \langle T_\psi f_n, (M-t)f_n\rangle + t\langle T_\phi f_n, f_n\rangle + t\langle T_\psi f_n, f_n\rangle| \\ &\geq |t|(\langle T_\phi f_n, f_n\rangle + \langle T_\psi f_n, f_n\rangle) - |\langle (M-t)f_n, T_\phi f_n\rangle + \langle T_\psi f_n, (M-t)f_n\rangle|. \end{aligned}$$

If $\lim_{n\to\infty} \langle T_\phi f_n, f_n\rangle > c$ or $\lim_{n\to\infty} \langle T_\psi f_n, f_n\rangle > c$, then $\lim_{n\to\infty} (M-t)f_n = 0$, it follows that $2c|t| > 2c|t|$ which is absurd. Therefore $\lim_{n\to\infty} \langle T_\phi f_n, f_n\rangle = c$ and $\lim_{n\to\infty} \langle T_\psi f_n, f_n\rangle = c$ and so $t = \lim_{n\to\infty} \langle M f_n, f_n\rangle = \lim_{n\to\infty} \langle T_\phi f_n, f_n\rangle - \lim_{n\to\infty} \langle T_\psi f_n, f_n\rangle = 0$. Hence $M = 0$. That implies $\phi = \psi$ which proves **(i)**.
Next assume that for some $1 \leq p < \infty$, we have $2c\|T_{\phi-\psi}\|_p = \|T_\phi^2 - T_\psi^2\|_p < \infty$. Then $2c\|M\|_p = \|T_\phi M + M T_\psi\|_p < \infty$. Since $M$ is a self-adjoint and also compact, it must be diagonalizable. Let $\{e_n\}_{n=0}^\infty$ be an orthonormal basis for $A^2(\mathbb{D})$ such that $Me_n = t_n e_n$. Then

$$\begin{aligned} (2c)^p \|M\|_p^p &= \|T_\phi M + M T_\psi\|_p^p \\ &\geq \sum_{n=0}^\infty |\langle (T_\phi M + M T_\psi) e_n, e_n\rangle|^p \\ &= \sum_{n=0}^\infty |t_n(\langle T_\phi e_n, e_n\rangle + \langle T_\psi e_n, e_n\rangle)|^p. \end{aligned}$$

If $\langle T_\phi e_n, e_n\rangle > c$ or $\langle T_\psi e_n, e_n\rangle > c$ for some $n$, then $(2c)^p\|M\|_p^p > (2c)^p\|M\|_p^p$ which is absurd. Thus $\langle T_\phi e_n, e_n\rangle = \langle T_\psi e_n, e_n\rangle = c$ for all $n$, and so $t_n = \langle M e_n, e_n\rangle = 0$ for all $n$, hence $M = 0$. That implies $\phi = \psi$ which proves **(ii)**. □

**Theorem 4.4** *Let $\phi, \psi \in L^{\infty}(\mathbb{D})$ with $\phi \geq 0$. Assume that $\sigma(S_{\psi}) \subset [0, \infty)$, $T_{\phi}S_{\psi} = S_{\psi^+}T_{\phi}$ and $T_{\phi}^2 S_{\psi} \geq 0$.*

*(i) If $\overline{Range\ T_{\overline{\phi}}} = A^2(\mathbb{D})$. Then $S_{\psi} \geq 0$.*
*(ii) If Kernel $S_{\psi}T_{\phi} = \{0\}$, then $S_{\psi} \geq 0$.*
*(iii) If $T_{\phi}$ is bounded below, then $S_{\psi} \geq 0$.*

***Proof*** **(i)**Since $\phi \geq 0$, $T_{\phi} \geq 0$. Further since $\sigma(S_{\psi}) \subset [0, \infty)$ and $T_{\phi}S_{\psi} = S_{\psi^+}T_{\phi}$, then it follows from [1] that $T_{\phi}S_{\psi}$ is positive. Therefore,

$$T_{\phi}S_{\psi} = (T_{\phi}S_{\psi})^* = S_{\psi^+}T_{\overline{\phi}}.$$

Thus,

$$\begin{aligned} T_{\phi}S_{\psi}T_{\overline{\phi}} &= S_{\psi^+}T_{\overline{\phi}}T_{\overline{\phi}} \\ &= (T_{\phi}^2 S_{\psi})^* \\ &= T_{\phi}^2 S_{\psi} \quad (\because T_{\phi}^2 S_{\psi} \geq 0). \end{aligned}$$

Hence,

$$\begin{aligned} \langle S_{\psi}T_{\overline{\phi}}f, T_{\overline{\phi}}f \rangle &= \langle T_{\phi}S_{\psi}T_{\overline{\phi}}f, f \rangle \\ &= \langle T_{\phi}^2 S_{\psi}f, f \rangle \geq 0, \quad f \in A^2(\mathbb{D}). \end{aligned}$$

So the hypothesis $\overline{Range\ T_{\overline{\phi}}} = A^2(\mathbb{D})$ implies that $S_{\psi} \geq 0$.

**(ii)** Since *Kernel* $T_{\phi} \subset$ *Kernel* $S_{\psi}T_{\phi} = \{0\}$, then $T_{\phi}$ is one-one. Therefore,

$$\overline{Range\ T_{\overline{\phi}}} = (Kernel\ T_{\overline{\phi}})^{\perp} = (Kernel\ T_{\phi})^{\perp} = \{0\}^{\perp} = A^2(\mathbb{D}).$$

Hence **(ii)** is proved by **(i)**.

**(iii)** Since $T_{\phi}$ is bounded below, then $T_{\phi}$ is one-one. That is *Kernel* $T_{\phi} = \{0\}$. Hence

$$\overline{Range\ T_{\overline{\phi}}} = \overline{Range\ T_{\phi}} = (Kernel\ T_{\phi})^{\perp} = \{0\}^{\perp} = A^2(\mathbb{D}).$$

Therefore, the proof follows from **(i)**. □

## References


1. S.J. Bernau, Products of positive operators. Proc. Camb. Phil. Soc. **76**, 415–416 (1974)
2. R. Bouldin, The product of operators with closed range. Tohoku Math. J. **25**, 359–363 (1973)
3. R.G. Douglas, On majorization, factorization and range inclusion of operators on Hilbert space. Proc. Am. Math. Soc. **17**, 413–415 (1966)
4. M. Gurdal, F. Sohret, Some results for Toeplitz operators on the Bergman space. Appl. Math. Comput. **218**, 789–793 (2011)

5. E. Heinz, On an inequality for linear operators in Hilbert space, in *Report on Operator Theory and Group Representations*, Publ. No. 387 (National Academy of Sciences-National Research Council, Washington, D.C., 1995), pp. 27–29
6. S. Ishikawa, Fixed points and iteration of a nonexpansive mapping in a Banach space. Proc. Am. Math. Soc. **59**, 65–71 (1976)
7. I.B. Jung, E. Ko, C. Pearcy, Aluthge transforms of operators. Integr. Equs. Oper. Theory **37**, 437–448 (2000)
8. T. Kato, Notes on some inequalities for linear operators. Math. Ann. **125**, 208–212 (1952)
9. C. Padhy, P.K. Jena, S.K. Paikray, Aluthge transform of operators on the Bergman space. Arab. J. Math. **8**, 1–11 (2019)
10. C. Padhy, P.K. Jena, S.K. Paikray, Operator matrices on the Bergman space. Filomat **36**(10), 3311–3320 (2022)
11. C. Padhy, P.K. Jena, S.K. Paikray , Unitariness of operators. Miskolc Math. Notes **22**(2), 831–840 (2021)
12. K. Zhu, *Operator Theory in Function Spaces, Monographs and Textbooks in pure and applied Mathematics*, vol. 139 (Marcel Dekker, New York, 1990)
13. A. Zygmund, *Trigonometric Series*, vol. I, II. (Cambridge University Press, Cambridge, New York, Melbourne, 1997)

# Fundamental Theorem of Approximation for Linear and Non-linear Sampling Kantorovich Operators

**Karunesh Kumar Singh and Digvijay Singh**

**Abstract** In this chapter, first we give the brief introduction of sampling theory and how it connects with the theory of sampling Kantorovich operators. Basically, our main task is to prove the fundamental theorem of approximation for different operators related to sampling Kantorovich operators. More preciously, the approximation theorem is proved for linear and non-linear operators using sampling Kantorovich theory. Eventually, we will illustrate an example for the justification of the fundamental theorem.

**Keywords** Orlicz space · Sampling Kantorovich operators · Order of moments · Fejer kernel

## 1 Introduction

Sampling theory is a branch of statistics that deals with the sampling of data from a population to make inferences about the population based on the characteristics observed in the sample whereas the Kantorovich operator, named after the mathematician Leonid Vitaliyevich Kantorovich, is an operator used in functional analysis and approximation theory. The main goal of sampling theory is to study how to obtain a representative subset (sample) of a larger population, and how to make valid statistical inferences about the population based on this sample while the Kantorovich operator is often employed in the study of approximation theory, where the goal is to find a function that closely approximates another function. It is particularly used in the field of numerical analysis to approximate functions and solve mathematical problems. Later on, the idea of the combined work of Kantorovich operators [5, 14] and sample theory came into the picture in the form of signal processing using the concept of band limited. This idea is given by Butzer and Stens [4].

K. K. Singh (✉) · D. Singh
Department of Applied Sciences and Humanities, Institute of Engineering and Technology, Lucknow 226021, Uttar Pradesh, India
e-mail: kksiitr.singh@gmail.com



S. A. Mohiuddine et al. (eds.), *Positive Operators and Fixed-Point Theorems with Applications*, Industrial and Applied Mathematics,
https://doi.org/10.1007/978-981-96-2476-8_9

Inspired by these works, we shall demonstrate the different forms of sampling Kantorovich operators with examples for the justification of the fundamental theorem of approximation. Finally, we give its application in image processing that how image changes when we increase its sampling rate.

## 2 Preliminaries

The idea of proving the fundamental theorem of approximation was proved with help of Weierstrass theorem which was firstly derived by the German mathematician Weierstrass in [15] which states as follows:

**Theorem 2.1** *For any continuous function $f : [a, b] \to \mathbb{R}$ and for any $\varepsilon > 0$, there exists a polynomial $P(x)$ such that for all $x$ in $[a, b]$:*

$$|f(x) - P(x)| < \varepsilon.$$

After that, motivated from this idea, many mathematicians have derived this theorem [11] like Runge (1885), Picard (1891), Lerch (1892 and 1903), Volterra (1897), Lebesgue (1898), Mittag-Leffler (1900), Fejér (1900 and 1916), Landau (1908), De la Vallée Poussin (1908), Jackson (1911), Sierpinski (1911), Bernstein (1912), Montel (1918), etc.

Subsequently, the researchers had extended this idea in terms of positive linear operators and explored their order of convergence deeply but here, we confine our work to understand the phenomena of constructing the positive linear operator in sampling theory. Therefore, in the following section, we will give the idea of construction of an operator in sampling theory. Firstly, let us define a function $f_s$ which is band limited in$[-w\pi, w\pi]$, where $w$ is positive real number then reconstruction of function can be done completely with the help of sample values $\frac{n}{w}, n \in \mathbb{Z}$. The value of $\frac{n}{w}$ denotes the node points, spaced equally in the following form:

$$f_s(x) = \sum_{n=-\infty}^{+\infty} f_s\left(\frac{n}{w}\right) sinc\pi(wx - n), \tag{2.1}$$

where the sinc function is defined as

$$\text{sinc}(x) = \begin{cases} \frac{\sin(x)}{x}, & \text{if } x \neq 0 \\ 1, & \text{if } x = 0. \end{cases}$$

The above result is known as sampling theorem derived by E. T. Whittaker (1915), V. A. Kotelǹikov (1933), and C. E. Shannon (1940/49) [12]. Due to some shortcomings regarding choosing the node points, researchers found the new way to get rid of from this problem by introducing a function (kernel function) which is given by the expression:

$$(\mathcal{S}_w^{\xi} f_s)(x) = \sum_{n=-\infty}^{+\infty} f_s\left(\frac{n}{w}\right) \xi(wx - n), \tag{2.2}$$

where $\xi : \mathcal{R} \to \mathcal{R}$ is called a Kernel function with the following properties:

- **K1**: $\xi \in L^1[\mathcal{R}]$ and bounded in $(-\epsilon, +\epsilon)$.
- **K2**: $\sum_{n=-\infty}^{+\infty} \xi(wx - n) = 1$.
- **K3**: $m_\kappa(\xi) = \sum_{n=-\infty}^{n=+\infty} \|(wx - n)\|_2^\kappa |\xi(wx - n)| < +\infty$.

The term $m_\kappa(\xi)$ denotes the moments of a function.

Besides that, there are some following additional interferences of kernel function that have been derived by Danilo Castarelli, Gianluca Vinti [8].

**Lemma 2.1** *Let $\xi : \mathcal{R} \to \mathcal{R}$ be a Kernel function then we have*

- *Moment of order zero is always finite, i.e., $m_0\xi(x) = \sup_{x\in\mathcal{R}} \sum_{n\in\mathcal{Z}} \xi(x - n) < +\infty$.*
- *For any $\rho > 0$,*

$$\lim_{w\to\infty} \sum_{\|(wx-n)\|\geq \rho w} \xi(wx - n) = 0,$$

*uniformly for all $x \in \mathcal{R}$.*
- *For any $\rho > 0$ and $\upsilon > 0$, $\exists \quad P > 0$ such that*

$$w \int_{\|x\|_2 > P} \xi(wx - n)dx < \upsilon$$

*as $w \to \infty$ and $n$ is taken in a such a way that $\|n\|_2 > \rho w$.*

***Note:*** *The generalized form $m_\kappa(\xi) = \sum_{n\in\mathcal{Z}} \|(n - wx)\|_2^\kappa |\xi(n - wx)|$ is known as moment of order $k$. Furthermore, if we take the* sup *over $x \in \bar{R}$ and denote it by $M_\kappa(\xi)$.*

Suppose $\mathcal{N}, \bar{\mathcal{N}}$ and $\mathcal{R}$ denotes the set of natural numbers, set of natural numbers including zero, and set of real numbers, respectively. Further, let us denote that $C(I), I \subset \mathcal{R}$ be the space of all uniformly continuous bounded function with respect to the supremum norm defined by $\|f\|_{C(I)} = \sup_{x\in I} |f(x)|$. Extensively, suppose $C^k(I) = \left\{f \in C(I) : \frac{d^k f}{dx^k} \in C(I), k \in \bar{\mathcal{N}}\right\}$, $C_0(I) = \{f \in C(I) : f \text{ is compact over } I\}$ and $C_0^k(I) = \left\{f \in C(I) : f^{(k)} \text{ is compact over } I\right\}$.

In similar fashion, let us denote some generalized notations. Suppose $\mathcal{N}^n$ be a set of vectors $\mathbf{a} = (a_1, a_2, \cdots, a_n)$ of $n$-tuples such that the $\sum_{i=1}^{n} \mathbf{a}_i = 1$ with each $a_i \in \mathcal{N}$ and $1 \leq i \leq n$. Similarly, the set of vectors for $\mathcal{N}_0^n$ and $\mathcal{Z}^n$ can be defined for non-negative integers and integers, respectively. Furthermore, $\mathcal{R}^n$ is defined in its usual manner as $(x_1, x_2, \cdots, x_n)$, $x_i \in \mathcal{R}$ with $1 \leq i \leq n$ and $\mathcal{R}_+^n$ denote vector of $n$-dimension with each $x_i > 0$. Let $\mathbf{x} = (x_1, x_2, \cdots, x_n)$ and $\mathbf{y} = (y_1, y_2, \cdots, y_n)$ be the vectors of $n$-dimension then we say $\mathbf{x} > \mathbf{y}$ if each $1 \leq i \leq n$, $x_i > y_i$. Moreover, the null and identity vectors for $n$-tuples are defined as follows:

$$\mathbf{0} = (0, 0, \cdots, 0)$$

$$\mathbf{1} = (1, 1, \cdots, 1),$$

respectively.

The algebraic properties of vectors of dimension $n$ are given as follows:

$$\mathbf{x} + \mathbf{y} = (x_1 + y_1, x_2 + y_2, \cdots, x_n + y_n);$$

$$\mathbf{kx} = (k_1 x_1, k_2 x_2, \cdots k_n x_n);$$

$$\frac{\mathbf{x}}{\mathbf{y}} = \left(\frac{x_1}{y_1}, \frac{x_2}{y_2}, \cdots \frac{x_n}{y_n}\right), y_i \neq 0, 1 \leq i \leq n;$$

$$k^{\mathbf{x}} = \left(k^{x_1}, k^{x_2}, \cdots, k^{x_n}\right), k > 0.$$

The norm of $\mathbf{x}$ is defined as follows: $\|\mathbf{x}\| = \sqrt{x_1^2 + x_2^2 + \cdots + x_n^2}$. The metric defined over the vectors is given by $d(x, y) = \|\mathbf{x} - \mathbf{y}\|$. Suppose $\mathbf{w} = (w_1, w_2, \cdots w_n) \in \bar{\mathcal{R}}^n$, we mean $\underline{w} = \min(w_1, w_2, \cdots, w_n) \to +\infty$.

## 3 Idea of Functional in Orlicz Space

Now, we will give some basic properties of functional in Orlicz space [16] and later on we prove the fundamental theorem of approximation using sampling Kantorovich operator in multivariate setting.

Let us consider a function $\psi : [0, \infty) \to [0, \infty)$ with the following assumptions:

(1) $\psi$ is continuous.
(2) $\psi$ is non-decreasing in $[0, \infty)$ with $\psi(0) = 0$ and $\psi(\mathbf{x}) > 0, \forall \mathbf{x} > 0$.
(3) We call it a $\psi$-function if $\lim_{\mathbf{x} \to \infty} \psi(\mathbf{x}) = +\infty$.

Further, assume that $M(R^n)$ denotes the space of all measurable (Lebesgue) functions on $R^n$. For fixed $\psi$-function and $\forall f \in M(R^n)$, we define a functional $J : M(R^n) \to [0, \infty)$ as follows:

$$J^{\psi}[f] = \int_{\mathcal{R}^n} \psi(|f(\mathbf{p})|)d\mathbf{p},$$

which is known as a modular functional on $M(\mathcal{R}^n)$. Eventually, Orlicz space is generated by the functional defined based on $\psi$-function. Suppose $L_1^{\psi}[\mathcal{R}^n]$ denotes an Orlicz space which is given as

$$L_1^{\psi}[\mathcal{R}^n] = \{f \in M(\mathcal{R}^n) : J^{\psi}[\Lambda f] < +\infty \text{ for some } \Lambda > 0\}.$$

Based on above space, let us define another space $L_2^{\psi}[\mathcal{R}^n]$ as follows:

$$L_2^{\psi}[\mathcal{R}^n] = \{f \in L_1^{\psi}[\mathcal{R}^n] : J^{\psi}[\Lambda f] < +\infty \text{ for some } \Lambda > 0\}.$$

Noting the fact that $C_0(\mathcal{R}^n) \subset L_2^{\psi}[\mathcal{R}^n]$ and $L_2^{\psi}[\mathcal{R}^n]$) is a vector subspace of all finite elements of $L_1^{\psi}[\mathcal{R}^n]$. Furthermore, these two spaces are the same if and only if $\psi$ satisfies the following condition:

$$\exists K > 0 \text{ such that } \psi(2\mathbf{x}) \leq K\psi(\mathbf{x}), \mathbf{x} \in \bar{\mathcal{R}}. \qquad (\Delta_2)$$

## 3.1 *Modular Convergence (Weak Convergence $L_1^{\psi}[\mathcal{R}^n]$)*

Now, we define the new term called "modular Convergence" with the help of $L_1^{\psi}[\mathcal{R}^n]$. Let us denote a term called net of a function as $(f_w)_{w>0} \subset L_1^{\psi}[\mathcal{R}^n]$, which is said to be modularly convergent to $(f_w)_{w>0}$ if $\lim_{w\to\infty} J^{\psi}[\Lambda(f_w - f_s)] = 0$ for some $\Lambda > 0$.

## 3.2 *Luxemburg Norm (Strong Notion of Convergence in $L_1^{\psi}[\mathcal{R}^n]$)*

Let us define a norm $\|.\|_{\psi}$ on $L_1^{\psi}[\mathcal{R}^n]$ is said to be Luxemburg norm if it holds the following equality:

$$\|f\|_{\psi} = \left\{\Lambda > 0 : \left[J^{\psi}\left(\frac{f_s}{\Lambda}\right)\right] \leq \Lambda\right\}.$$

Subsequently, if $\psi$ is a convex function then we have the following quality:

$$\|f\|_{\psi} = \left\{\Lambda > 0 : \left[J^{\psi}\left(\frac{f_s}{\Lambda}\right)\right] \leq 1\right\}.$$

Conclusively, every Luxemburg norm convergence is a modular convergence but the converse need not be true, whereas the converse part will be true whenever $\psi$ has to satisfy the $\Delta_2-$ condition.

## 4 Main Results

Now, let us derive the main theorem of fundamental theorem of approximation for different forms of operators.

**Theorem 4.1** *Let $f_s \in C_0(I), I \subset \mathcal{R}, \forall x \in \mathcal{R}$. It remains to be a continuous function. Further, let us define the set of a positive linear operators $(\mathcal{S}_w^{\xi} f_s)_0(x)_{w>0}$ as defined in (2.2) which maps $C(I)$ to itself, then*

$$\lim_{w \to +\infty} (\mathcal{S}_w^{\xi} f_s)_0(x) = f_s(x).$$

***Proof*** Obviously, we have the following estimation as

$$|(\mathcal{S}_w^{\xi} f_s)(x) - f_s(x)| \leq \left( \sum_{|(wx-n)|<\rho w} + \sum_{|(wx-n)|\geq \rho w} \right) \left| \left( f_s(x) - f_s\left(\frac{n}{w}\right) \right) \right| |\xi(wx-n)|$$

$$\leq K_1 + K_2, \text{(say)}.$$

Since $f \in C_0(I), I \subset \mathcal{R}, \forall x \in \mathcal{R}$ then for $\upsilon > 0$ there exist $\rho$ such that $\left|\left(f_s - f_s\left(\frac{n}{w}\right)\right)\right| \leq \upsilon$ whenever $\left|x - \frac{n}{w}\right| \leq \rho$ with the upper bound $K_1 < \upsilon \|\mathcal{K}_w^{\xi}\|_{C(I)}$. As for concern of $K_2$, it is equal to zero because of the compact support of $\xi$ in the interval $[-\rho w, \rho w]$. Hence, we have the required result. □

Now, we will prove the above theorem when the sampling values are taken exponentially as follows:

$$(\mathcal{S}_w^{\xi} f_s)_1(x) = \sum_{n \in \mathcal{Z}} \xi(e^{-n} x^w) f(e^{\frac{n}{w}}). \tag{4.1}$$

Essentially, let us denote $\mathcal{M}(\bar{\mathcal{R}})$ be the set of all Lebesgue measurable functions and $\mathcal{L}^{\infty}(\bar{\mathcal{R}})$ be the set of all bounded functions over $(0, \infty)$. For $1 \leq p < \infty, \mathcal{L}^p(\bar{\mathcal{R}})$ be a class of Lebesgue measurable and $p$-integrable functions. In order to derive the following theorem, let us define a complex valued function $f$ over $R^+$ is called log uniformly continuous function if for $\upsilon > 0$ there exists $\rho > 0$ such that $|f(u_1) - f(u_2)| \leq \upsilon$ whenever $|\log u_1 - \log u_2| \leq \rho, \forall u_1, u_2 \in \bar{\mathcal{R}}$.

**Theorem 4.2** *Suppose $f_s \in \mathcal{M}(\bar{\mathcal{R}}) \cap \mathcal{L}^{\infty}(\bar{\mathcal{R}}), \forall x \in \bar{\mathcal{R}}$, it remains to be continuous and the series (4.1)converges to $f_s(x)$ uniformly $\forall x \in \bar{\mathcal{R}}$ then we have*

$$\lim_{w \to \infty} (\mathcal{S}_w^{\xi} f_s)_1(x) = f(x),$$

*where $f, \xi : \bar{\mathcal{R}} \to \mathcal{R}$, $f \in L^{\infty}(\bar{\mathcal{R}})$ and $\xi$ satisfies following properties:*

*(1)* $\sum_{n \in \mathcal{Z}} \xi(e^{-v} x^{n}) = 1.$

*(2)* $m_{\kappa}(\xi) = \sum_{\mathcal{Z}} \|(e^{-n} x^{w})\|_{2}^{\kappa} |\xi(e^{-n} x^{w})| < +\infty$. *Further, if we take the* sup *over $x \in \bar{R}$, let us say it is denoted by $M_{\kappa}(\xi)$.*

*(3) For any $\rho > 0$,*

$$\lim_{w \to \infty} \sum_{|(n - w \log x)| > \rho w} \xi(e^{-n} x^{w}) = 0,$$

*uniformly for all $x \in \mathcal{R}$.*

***Proof*** As given that $f_s$ is continuous for each $x \in \bar{\mathcal{R}}$ and from the definition of uniformly continuity in log sense, we have

$$|(\mathcal{S}_{w}^{\xi} f_s)_1(x) - f_s(x)| \leq \left( \sum_{|(w \log x - n)| \leq \rho w} + \sum_{|(w \log x - n)| > \rho w} \right) \left|(f_s(x) - f_s\left(e^{\frac{n}{w}}\right)\right| |\xi(e^{-v} x^{n})|$$

$$\leq M_0(\xi) + 2\|f\|_{\infty} \sum_{|(n - w \log x)| > \rho w} \xi(e^{-v} x^{n}).$$

Clearly, the R.H.S. of above inequality goes to zero, since we choose $w \to +\infty$ and $v \to 0$, which implies that

$$(\mathcal{S}_{w}^{\xi} f_s)(x)_1 \rightrightarrows f_s(x), \forall x \in (0, \infty).$$

□

Generally, a reconstruction formula for the band-limited functions are derived at a sequence of equally and unequally sample points over real line. Furthermore, another manipulation was done by Carlo Bardaro et al.[1], in which they have taken the mean sample values as given in the next theorem while dealing with the Kantorovich operators in the following expression:

$$(\mathcal{K} f_s)_2(x) = \sum_{n=-\infty}^{+\infty} \left( w \int_{\frac{k}{w}}^{\frac{k+1}{w}} f_s(p) dp \right) \xi(wx - n), \tag{4.2}$$

where $f_s, \xi : \mathcal{R} \to \mathcal{R}$ are defined as in Theorem 4.2.

**Theorem 4.3** *Suppose $f \in \mathcal{M}(\mathcal{R}) \cap \mathcal{L}^{\infty}(\mathcal{R}), \forall x \in \mathcal{R}$, it remains to be continuous and the series (4.2) converges to $f_s(x)$ uniformly $\forall x \in \mathcal{R}$ then we have*

$$\lim_{w \to \infty} (\mathcal{K}_{w}^{\xi} f_s)_2(x) = f_s(x),$$

*where $\xi : \mathcal{R} \to \mathcal{R}$, $f_s$ belongs from the set of all bounded functions and the properties of $\xi$ are mentioned in (2.2).*

***Proof*** Given that $f \in \mathcal{M}(\mathcal{R}) \cap \mathcal{L}^{\infty}(\mathcal{R})$, $\forall x \in \mathcal{R}$. Therefore, for $\upsilon > 0$, there exists $\rho > 0$ such that $|f(x_1) - f(x_2)| \le \upsilon$ whenever $|x_1 - x_2| \le \rho$. Setting $w > 0$ such that $\frac{1}{w} < \frac{\upsilon}{2}$, we have

$$|(\mathcal{K}_w^{\xi} f_s)_2(x) - f_s(x)| \le \left( \sum_{|(wx-n)| \le \rho w} + \sum_{|(wx-n)| > \rho w} \right) w \int_{\frac{k}{w}}^{\frac{k+1}{w}} |(f_s(p) - f_s(x)| \, |\xi(wx-n)|$$

$$\le K + K'(\text{ say}).$$

Let us set $p \in \left[\frac{k}{w}, \frac{k+1}{w}\right]$ whenever $(wx - n) < \frac{w\upsilon}{2}$.

$$|p - x| \le \left| u - \frac{1}{w} \right| + \left| \frac{1}{w} - x \right| < \frac{1}{w} + \frac{\upsilon}{2}.$$

$$K \le \sum_{|(wx-n)| \le \rho w} w \int_{\frac{k}{w}}^{\frac{k+1}{w}} |(f_s(p) - f_s(x))| \, |\xi(wx-n)| \le \upsilon m_{\beta, \prod^n}(\xi).$$

Now,

$$K' \le 2\|f\|_{\infty} \sum_{|(n - w \log x)| > \frac{\rho w}{2}} |\xi(wx - n)|.$$

Letting $\upsilon \to 0$ and $w \to +\infty$ which proves that $(\mathcal{K}_w^{\xi} f_s)_2(x)$ converges uniformly to $f_s(x)$. □

Consequently, many researchers did some sort of manipulation in the above-mentioned operators in sampling theory and extended these operators in $n$-dimensional space either. Therefore, the proofs are not so tedious work, but we will state the theorem and left the proofs for the readers.

Now, we state the following sampling Kantorovich operators in generalized version:

$$(\mathcal{K} f_s)_3(x) == \sum_{\mathbf{k}=-\infty}^{+\infty} \left( \frac{w^n}{A_{\mathbf{k}}} \int_{R_{\mathbf{k}}^w} f(\mathbf{p}) d\mathbf{p} \right) \xi(\mathbf{wx} - \mathbf{k}), \mathbf{x} \in \mathcal{R}^n, \tag{4.3}$$

where bold lowercase letters represent the vectors in n-tuples.

$$R_{\mathbf{k}}^w = \left[ \frac{k_1}{w}, \frac{k_1 + 1}{w} \right] \times \left[ \frac{k_2}{w}, \frac{k_2 + 1}{w} \right] \times \cdots \times \left[ \frac{k_n}{w}, \frac{k_n + 1}{w} \right], w > 0.$$

Noting the fact that the Lebesgue measure of $R_{\mathbf{k}}^w$ is $\frac{w^n}{A_{\mathbf{k}}}$, where $A_{\mathbf{k}} = \prod_{i=1}^n (k_{i+1} - k_i), i = 1, 2, \cdots, n$ and for the uniform distribution of sampling Kantorovich operators we assume $k_{k'} = k', k' \in \mathcal{N}$ with the properties of kernel function $\xi : \mathcal{R}^n \to \mathcal{R}$ as follows:

(i) $\xi(w\mathbf{x} - \mathbf{k}_{\mathbf{k}'}) \in L^1(\mathcal{R}^n)$, $\mathbf{x} \in \mathcal{R}^n$, $\mathbf{p} \in \mathcal{R}^n$, and $w > 0$; is bounded in a neighborhood of $\mathbf{0}$.
(ii) $\sum_{\mathbf{k}\in\mathcal{Z}^n} \xi(w\mathbf{x} - \mathbf{k}_{\mathbf{k}'}) = 1$.
(iii) $m_{\beta,\prod^n}(\xi) = \sup_{\mathbf{x}\in\mathcal{Z}^n} \sum_{\mathbf{k}\in\mathcal{Z}^n} |\xi(w\mathbf{x} - \mathbf{k}_{\mathbf{k}'})| \|w\mathbf{x} - \mathbf{k}_{\mathbf{k}'}\|_2^{\beta} < +\infty$.

Again, noting the fact that $m_{\beta,\prod^n}(\xi)$ is the generalized moment of order $\beta$.

**Theorem 4.4** *Suppose* $f_s \in \mathcal{M}(\mathcal{R}) \cap \mathcal{L}^{\infty}(\mathcal{R})$, $\forall x \in \mathcal{R}$, *it remains to be continuous and the series (4.3) converges to* $f_s(x)$ *uniformly,* $\forall x \in \mathcal{R}$ *then we have*

$$\lim_{w\to\infty} (\mathcal{K}_w^{\xi} f_s)_3(x) = f_s(x),$$

*where* $\xi : \mathcal{R}^n \to \mathcal{R}$.

Similarly, we can extend this theorem by taking the sample values exponentially as mentioned in the following expression and proceed its proof with the help of Theorems 4.2 and 4.3.

$$(\mathcal{K} f_s)_4(x) = \sum_{n=-\infty}^{+\infty} \left( w \int_{\frac{k}{w}}^{\frac{k+1}{w}} f(e^p) dp \right) \xi\left(e^{-n} x^w\right), \tag{4.4}$$

where $f_s, \xi : \bar{\mathcal{R}} \to \mathcal{R}$ are defined in such a way that $f_s \in L^{\infty}(\bar{\mathcal{R}})$ and the properties of $\xi$ as in Theorem 4.2.

**Theorem 4.5** *Suppose* $f_s \in \mathcal{M}(\bar{\mathcal{R}}) \cap \mathcal{L}^{\infty}(\bar{\mathcal{R}})$, $\forall x \in \bar{\mathcal{R}}$, *it remains to be continuous and the series (4.4) converges uniformly to* $f_s(x)$ $\forall x \in \bar{\mathcal{R}}$ *then we have*

$$\lim_{w\to\infty} (\mathcal{K}_w^{\xi} f_s)_4(x) = f_s(x),$$

*where* $f_s, \xi : \bar{\mathcal{R}} \to \mathcal{R}$, $f_s \in L^{\infty}(\bar{\mathcal{R}})$ *and the properties of kernel function* $\xi$ *can be easily guessed from Theorem 4.2.*

**Remark 4.1** Similarly, we can prove that generalized sampling theorem for exponential sample values.

Finally, we give the brief introduction of non-linear sampling Kantorovich positive operators. In this context, Bardaro, Matellin, and many other researchers gave some important estimation based on the function $g : \mathcal{N} \to \mathcal{R}$ for the following operator by doing some sort modification. For more details, one can go through the article [2, 3, 6, 7, 9, 10]. Eventually, this type of operators are quit useful while studying fractional calculus in terms of sampling Kantorovich operators

$$(S_w^{\xi} f)(\mathbf{x}) = \sum_{\mathbf{k}=-\infty}^{+\infty} \left( < g(w) > \int_{R_{\mathbf{k}}^w} f(\mathbf{p}) d\mathbf{p} \right) \xi(g(w)\mathbf{x} - \mathbf{k}), \mathbf{x} \in \mathcal{R}^n,$$

where bold lowercase letters represent the vectors in n-tuples.

The idea of restricting the weight values only to natural numbers had been beautifully strategized in [13] by Gianluca Vinti and Luca Zampogni for generalized non-linear sampling Kantorovich which is given as follows:

$$(S^{\xi}_{g(w)} f)_5(\mathbf{x}) = \sum_{\mathbf{k}\in\mathbb{Z}^n} \xi\left(g(w)\mathbf{x} - \mathbf{x}_\mathbf{k}, \frac{g(w)}{A_k^{\alpha}} \int_{R^w_{\mathbf{x}_\mathbf{k}}} f(\mathbf{p})\, d\mathbf{p}\right) \tag{4.5}$$

$$R^w_{\mathbf{x}_\mathbf{k}} = \left[\frac{\mathbf{x}_{k_1}}{w}, \frac{\mathbf{x}_{k_1}+1}{w}\right] \times \left[\frac{\mathbf{x}_{k_2}}{w}, \frac{\mathbf{x}_{k_2}+1}{w}\right] \times \cdots \times \left[\frac{\mathbf{x}_{k_n}}{w}, \frac{\mathbf{x}_{k_n+1}}{w}\right], w > 0.$$

Noting the fact that the Lebesgue measure of $R^w_{\mathbf{x}_\mathbf{k}}$ is $\frac{A_{\mathbf{x}_\mathbf{k}}}{w^n}$, where $A_{\mathbf{x}_\mathbf{k}} = \prod_{i=1}^n (\mathbf{x}_{k_{i+1}} - \mathbf{x}_{k_i})$, $i = 1, 2, \cdots, n$. Suppose $\Pi^n = (x_\mathbf{k})_{\mathbf{k}\in Z^n}$ is a sequence with $-\infty < x_{\mathbf{k_i}} < x_{\mathbf{k_{i+1}}} < +\infty, i = 1, 2, \cdots, n$ such that $\lim_{k_i \to +,-\infty} x_\mathbf{k} = +, -\infty$. Moreover there exist positive numbers $\delta$, and $\Delta$ such that $\delta \le x_{\mathbf{k_{i+1}}} - x_{\mathbf{k_i}} \le \Delta$. Based on the above observation, we have the following inequality

$$\frac{\delta^n}{w^n} \le \frac{A_k}{w^n} \le \frac{\Delta^n}{w^n}.$$

## 4.1 Properties of a Kernel Function

Suppose $\Pi^n = (x_\mathbf{k})_{\mathbf{k}\in Z^n}$ is a sequence with $-\infty < x_{\mathbf{k_i}} < x_{\mathbf{k_{i+1}}} < +\infty, i = 1, 2, \cdots, n$ such that $\lim_{k_i \to +,-\infty} x_{\mathbf{k_i}} = +, -\infty$. Let us define a function $\xi : \mathcal{R}^n \times \mathcal{R} \to \mathcal{R}$, known as Kernel function with the following properties:

- $\mathbf{k}'_\mathbf{k} \mapsto \xi(w\mathbf{x} - \mathbf{x}_{\mathbf{k}'}, \mathbf{p}) \in l^1(Z^n)$, $\mathbf{x}_{\mathbf{k}'} \in \mathcal{R}^n$, $\mathbf{p} \in \mathcal{R}^n$ and $w > 0$;
- $\xi(w\mathbf{x}, 0) = 0$, for every $\mathbf{x} \in \mathcal{R}^n$
- $\xi$ is an $(L, \xi)$-Lipschitz kernel, that is, there exists a measurable function $L : \mathcal{R}^n \to \mathcal{R}_0^+$ such that $|\xi(w\mathbf{x}, \mathbf{p}) - \xi(w\mathbf{x}, \mathbf{v})| \le L(w\mathbf{x})|\xi(\mathbf{p} - \mathbf{v})|$, $\mathbf{x} \in \mathcal{R}^n$, $\mathbf{p}, \mathbf{v} \in \mathcal{R}^n$ and $w > 0$;
- $\forall j \in \mathcal{N}^+$ and $w > 0$, $\mathcal{T}_j^w(\mathbf{x}) = \sup_{\frac{1}{j} \le |\mathbf{p}| \le j} \sum_{\mathbf{k}'\in\mathcal{Z}^n} |\xi(w\mathbf{x}_{\mathbf{k}'}, \mathbf{p}) - 1| \to 0$, as $w \to +\infty$ uniformly w.r.t. $\mathbf{x} \in \mathcal{R}^n$.

If the function $\xi$ satisfies the above properties then we define a class of function denoted by H. Moreover, we assume that the function L satisfies the following assumptions:

(1) $L \in L^1(\mathcal{R}^n)$ and is bounded in a neighborhood of $0 \in \mathcal{R}^n$.
(2) $\mathbf{k}' \mapsto L(\mathbf{u} - \mathbf{x}_{\mathbf{k}'}) \in l^1(\mathcal{Z}^n)$ for every $\mathbf{x} \in \mathcal{R}^n$.

(3) There exist a point $\beta > 0$ such that $m_{\beta,\prod^n}(L) = \sum_{\mathbf{k}=-\infty}^{\mathbf{k}=+\infty} L(\mathbf{u} - \mathbf{x}_{\mathbf{x}_\mathbf{k}})\|\mathbf{u} - \mathbf{x}_{\mathbf{x}_\mathbf{k}}\|_2^\beta < +\infty$. If the function L satisfies the above condition then we define a class say $\mathcal{L}$.

**Note 1:** Instead of assuming that the function L is bounded in a neighborhood of $0 \in \mathcal{R}^n$ and that $m_{\beta,\prod^n}(L) < +\infty$.

**Lemma 4.1** *Let $\xi$ be a kernel and L be defined as above. Then*

*(1)* $m_{\beta,\prod^n}(L) = \sup_{x \in \mathcal{R}^n} \sum_{k=-\infty}^{k=+\infty} L(\boldsymbol{u} - \boldsymbol{x}_k) < +\infty$.
*(2)* $\forall \delta > 0$,

$$\sum_{\|g(w)\boldsymbol{x}-\boldsymbol{x}_{x_k}\|>\rho g(w)} |L(g(w)\boldsymbol{x} - \boldsymbol{x}_k)| = 0,$$

*uniformly with respect to* $\boldsymbol{x} \in \mathcal{R}^n$.
*(3) For every* $\upsilon, \rho > 0$, *there exist a positive real number M*

$$\int_{\|\boldsymbol{x}\|_2>M} \left|w^n L(g(w)\boldsymbol{t} - \boldsymbol{x}_k)\right| < \upsilon,$$

*with* $\|\boldsymbol{x}\|_2 \le \delta g(w)$ *as* $\boldsymbol{x}, w \to +\infty$.

**Remark 4.2** Noting the fact that $(\mathcal{K}^{\xi}_{g(w)} f)_5(\mathbf{x}) \in L^\infty(\mathcal{R}^n)$ is well defined from the following inequality:

$$|(\mathcal{K}^{\xi}_{g(w)} f)_5(\mathbf{x})| \le \left|\sum_{\mathbf{k}\in\mathbb{Z}^n} \xi\left(g(w)\mathbf{x} - \mathbf{x}_\mathbf{k}, \frac{g(w)}{A_k^\alpha}\int_{R^w_{\mathbf{x}_\mathbf{k}}} f(\mathbf{p})\,d\mathbf{p}\right)\right|.$$

$$|(\mathcal{K}^{\xi}_{g(w)} f)_5(\mathbf{x})| \le \sum_{\mathbf{x}_\mathbf{k}\in Z^n} L(g(w)\mathbf{x} - \mathbf{x}_\mathbf{k}) \left|\frac{g(w)}{A_k^\alpha}\int_{R^w_{\mathbf{x}_\mathbf{k}}} f(\mathbf{p})\,d\mathbf{p}\right|$$

$$\le m_{0,\prod^n}(L) \times \frac{\|f\|_\infty}{A_k^\alpha} < \infty,$$

for every $\mathbf{x} \in \mathcal{R}^n$ and $w > 0$. From the above expression, we conclude that $(S^{\xi}_{g(w)} f)_5 : C(\mathcal{R}) \to L^\infty(\mathcal{R})$.

**Theorem 4.6** *Let f be a bounded function such that* $\forall \boldsymbol{x} \in \mathcal{R}^n$ *it remains to be a continuous function. Further, suppose that* $\xi \in \mathcal{H}$ *then*

$$\lim_{g(w)\to+\infty} (\mathcal{K}^{\xi}_{g(w),\alpha} f)_5(\boldsymbol{x}) = f(\boldsymbol{x}), \forall \boldsymbol{x} \in \mathcal{R}^n.$$

***Proof*** As given that $f$ is continuous forall $\mathbf{x} \in \mathcal{R}^n$ then we can set the inequality as follows.

$$\begin{aligned}
&\left|(\mathcal{K}^{\xi}_{g(w)}f)_5(\mathbf{x}) - f(\mathbf{x})\right| \\
&\leq \sum_{\mathbf{k}\in Z^n}\left|\xi\left(g(w)\mathbf{x}-\mathbf{x_k}, \frac{g(w)}{A_k}\int_{\frac{\mathbf{k}}{g(w)}}^{\frac{\mathbf{k+1}}{g(w)}} f(\mathbf{p})d(\mathbf{p})\right) - \xi\left(g(w)\mathbf{x}-\mathbf{x_k}, f(\mathbf{x})\right)\right| \\
&+ \left|\sum_{\mathbf{k}\in Z^n}\xi\left(g(w)\mathbf{x}-\mathbf{x_k}, f(\mathbf{x})\right) - f(\mathbf{x})\right|
\end{aligned}$$

$$\begin{aligned}
&\leq \sum_{\mathbf{k}\in Z^n} L(g(w)\mathbf{x}-\mathbf{x_k})\xi\left(\frac{g(w)}{A_k}\int_{\frac{\mathbf{k}}{g(w)}}^{\frac{\mathbf{k+1}}{g(w)}} |f(\mathbf{u}) - f(\mathbf{x})|d(\mathbf{u})\right) \\
&+ |\sum_{\mathbf{x}\in Z^n}\xi\left(g(w)\mathbf{x}-\mathbf{x_k}, f(\mathbf{x})\right) - f(\mathbf{x})| := I_1 + I_2, (say).
\end{aligned}$$

By definition of continuity, for a fixed $\upsilon > 0$, $\exists\rho > 0$ such that $|f(\mathbf{p}) - f(\mathbf{x})| < \upsilon$, whenever $\|\mathbf{p}-\mathbf{x}\|_2 < \rho$, we have

$$\left|(\mathcal{K}^{\xi}_{g(w)}f)_5(\mathbf{x}) - f(\mathbf{x})\right| = \left\{\sum_{\|g(w)\mathbf{x}-\mathbf{x_k}\|_2\leq\frac{\delta g(w)}{2}} + \sum_{\|g(w)\mathbf{t}-\mathbf{x_k}\|_2>\frac{\delta g(w)}{2}}\right\} L(g(w)\mathbf{x}-\mathbf{x_k})$$

$$\times\,\xi\left(\frac{g(w)}{A_k}\int_{\frac{\mathbf{k}}{g(w)}}^{\frac{\mathbf{k+1}}{g(w)}} |f(\mathbf{u}) - f(\mathbf{x})|d(\mathbf{u})\right) := I_1' + I_2'(say).$$

Now, $\|\mathbf{u}-\mathbf{x}\|_2 \leq \frac{\delta g(w)}{2} \implies \|\mathbf{u}-\mathbf{x}\|_2 \leq \|\mathbf{u} - \frac{\mathbf{x_k}}{g(w)}\|_2 + \|\mathbf{x} - \frac{\mathbf{x_k}}{g(w)}\|_2 = \sqrt{n}\times\frac{\Delta}{g(w)} + \frac{\delta}{2}$
For a sufficiently large value of $g(w) > 0$, $\sqrt{n}\times\frac{\Delta}{g(w)} < \frac{\delta}{2}$.

$$\begin{aligned}
I_1' &= \sum_{\|\mathbf{u}-\mathbf{x}\|_2\leq\frac{\delta g(w)}{2}} L(g(w)\mathbf{x}-\mathbf{x_k})\xi\left(\frac{g(w)}{A_k}\int_{\frac{\mathbf{k}}{g(w)}}^{\frac{\mathbf{k+1}}{g(w)}} |f(\mathbf{u}) - f(\mathbf{x})|d\mathbf{u}\right) \\
&= \frac{\xi(\epsilon)m_{0,\prod^n}(L)}{A_k}.
\end{aligned}$$

*Now*,

$$I_2' = \sum_{\|\mathbf{u}-\mathbf{x}\|_2>\frac{\delta g(w)}{2}} L(g(w)\mathbf{x}-\mathbf{x_k})\xi\left(\frac{g(w)}{A_k}\int_{\frac{\mathbf{k}}{g(w)}}^{\frac{\mathbf{k+1}}{g(w)}} |f(\mathbf{u}) - f(\mathbf{x})|d\mathbf{u}\right).$$

For $w \to \infty$ and using the property (2) of the lemma 2.1, we have

$$I_2^{'} \leq \sum_{\|\mathbf{u}-\mathbf{x}\|_2 > \frac{\delta g(w)}{2}} L(g(w))\mathbf{t} - \mathbf{x_k})\xi\left(\frac{2\|f\|_\infty}{A_\mathbf{k}}\right) \to \mathbf{0}.$$

Now, let us find the estimation for the second term $I_2$. Since $f$ is bounded so there exists a number $j \in \mathbf{N}$ such that $|f(\mathbf{x})| \leq j$ with $\frac{1}{j} < 0$. Setting $D_j = \{\mathbf{x} \in \mathbf{R}^n : 0 \leq f(\mathbf{x}) \leq \frac{1}{j}\}$ and introducing the characteristic function $\mathcal{X}_{D_j}$ in the following way:

$$I_2 = \sum_{\mathbf{k}\in Z^n} |\xi\left(g(w)\mathbf{x} - \mathbf{x_k}, f(\mathbf{x})\mathcal{X}_{D_j}\right) - f(\mathbf{x})\mathcal{X}_{D_j} + \xi\left(g(w)\mathbf{x} - \mathbf{x_k}, f(\mathbf{x})\mathcal{X}_{\mathbf{R}^n - D_j}\right) - f(\mathbf{x})\mathcal{X}_{\mathbf{R}^n - D_j}|$$

$$\leq \sum_{\mathbf{k}\in Z^n} |\xi\left(g(w)\mathbf{t} - \mathbf{x_k}, f(\mathbf{x})\mathcal{X}_{D_j}\right) - f(\mathbf{x})\mathcal{X}_{D_j}| + |f(\mathbf{x})|\mathcal{T}^{\alpha}_{g(w)}(\mathbf{t})$$

$$I_2 := I_{2,a_1} + I_{2,a_2}(\text{say})$$

$$I_{2,a_1} = \sum_{\mathbf{x_k}\in Z^n} L\left(g(w)\mathbf{t} - \mathbf{x_k}\right)\xi\left|(f(\mathbf{x})\mathcal{X}_{D_j})\right| + \left|(f(\mathbf{x})\mathcal{X}_{D_j})\right| \leq \xi(\frac{1}{j})m_{0,\prod^n}(L) + \epsilon.$$

Letting $\epsilon \to \mathbf{0}$, we have $I_{2,a_1} = 0$ and using the last property of $\xi$, we have $I_{2,a_2} = 0$. Thus, we have proved that $I_1 = 0$ and $I_2 = 0$. Eventually, we got the required result. □

## 5 Illustration with Graphical Representation

Now, we will illustrate an example for the justification of the fundamental theorem of approximation. Let us take the sampling series as mentioned in (4.2) by taking the equation $< x_k >= k$ which is given as follows:

$$(\mathcal{S}f_s)_5(x) = \sum_{n=-\infty}^{+\infty}\left(w\int_{\frac{k}{w}}^{\frac{k+1}{w}} f(p)dp\right)F(wx - n),$$

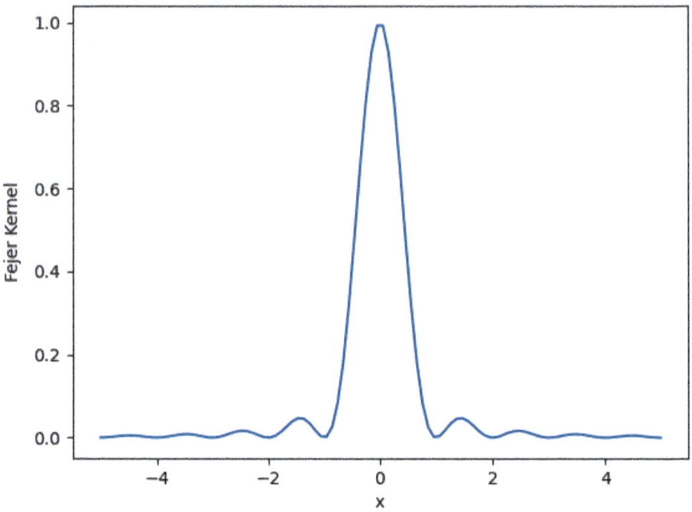


**Fig. 1** Fejer Kernel

where $F(x)$ is known as Kernel function with $F(x) = \frac{1}{2} sinc^2\left(\frac{x}{2}\right)$. Further assume that if $\hat{f}(p)$ is the Fourier transformation of $f(x)$ in $L^2(R)$ sense then $\mathcal{F}(sincx)$ is given by

$$\hat{f}(p) = \begin{cases} 1 - \left|\frac{p}{\pi}\right| & \text{if } -\pi \le p \le \pi \\ 0 & \text{if otherwise.} \end{cases}$$

Obviously, all the properties of kernel functions as mentioned in the (2.2) are satisfied. Pictorially, the kernel function can be represented in Fig. 1.

Additionally, let us define a continuous function which is given by

$$f_s(x) = \begin{cases} 0.40 & \text{if } -5 \le x < -2.5 \\ 0.75 & \text{if } -2.5 \le x < 0 \\ 0.675 & \text{if } 0 \le x < 2.5 \\ 0.625 & \text{otherwise.} \end{cases}$$

Now, in the final step using Python 3.11.7, we evaluate $(\mathcal{K} f_s)(x)$ for different values of $w = 2, 6, 10$ in increasing order. As shown in Fig. 2, we can easily observe that if we increase the value of $w$ in $(\mathcal{K} f_s)(x)$, we see that it tends to the value of $f_s(x)$ which is nothing but the statement of fundamental theorem of approximation.

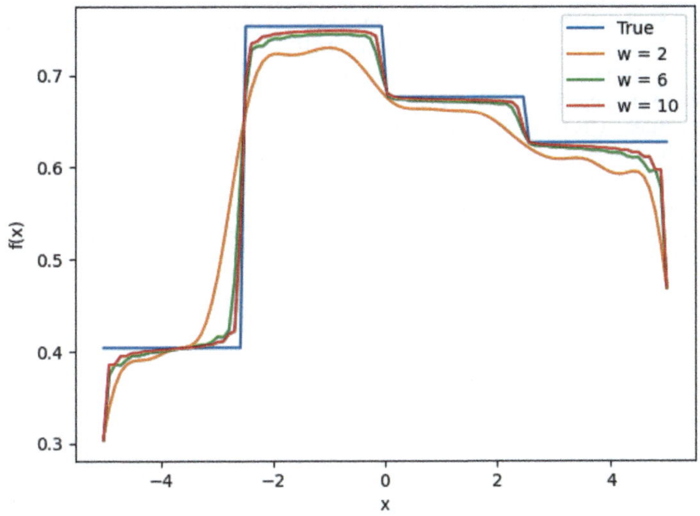


**Fig. 2** Comparison between $(\mathcal{S} f_s)(x)$ and $f_s(x)$ for different values of $w$

# References


1. C. Bardaro, G. Vinti, P.L. Butzer, R.L. Stens, Kantorovich-type generalized sampling series in the setting of Orlicz spaces, in *Sampling Theory in Signal and Image Processing*, vol. 6, pp. 29–52 (2007)
2. C. Bardaro, G. Vinti, H. Karsli, Nonlinear integral operators with homogeneous kernels: pointwise approximation theorems. Appl. Anal. **90**(3–4), 463–474 (2011)
3. C. Bardaro, I. Mantellini, On pointwise approximation properties of multivariate semi-discrete sampling type operators. Res. Math. **72**, 1449–1472 (2017)
4. P.L. Butzer, R.L. Stens, Sampling theory for not necessarily band-limited functions: a historical overview. SIAM Rev. **34**(1), 40–53 (1992)
5. P.L. Butzer, S. Ries, R.L. Stens, Approximation of continuous and discontinuous functions by generalized sampling series. J. Approx. Theory **50**(1), 25–39 (1987)
6. N. Cetin, D. Costarelli, G. Vinti, Quantitative estimates for nonlinear sampling Kantorovich operators. Res. Math. **76**(2), 80 (2021)
7. D. Costarelli, G. Vinti, Approximation by nonlinear multivariate sampling Kantorovich type operators and applications to image processing. Numer. Funct. Anal. Optim. **34**(8), 819–844 (2013)
8. D. Costarelli, G. Vinti, Approximation by multivariate generalized sampling Kantorovich operators in the setting of Orlicz spaces. Bollettino dell'Unione Matematica Italiana **4**(3), 445–468 (2011)
9. D. Costarelli, G. Vinti, Rate of approximation for multivariate sampling Kantorovich operators on some functions spaces, pp. 455–481 (2014)
10. D. Costarelli, G. Vinti, Approximation results by multivariate sampling Kantorovich series in Musielak-Orlicz spaces. Dolomites Res. Notes Approx. DRNA **12**, no. 12.1, 7–16 (2019)
11. V. Gupta, R.P. Agarwal, *Convergence Estimates in Approximation Theory*, vol. 13 (Springer, Cham, 2014)
12. R.J. Marks, II (ed.), Advanced Topics in Shannon Sampling and Interpolation Theory. Springer Science and Business Media (2012)

13. G. Vinti, L. Zampogni, Approximation by means of nonlinear Kantorovich sampling type operators in Orlicz spaces. J. Approx. Theory **161**(2), 511–528 (2009)
14. G. Vinti, L. Zampogni, A unifying approach to convergence of linear sampling type operators in Orlicz spaces, pp. 573–600 (2011)
15. K. Weierstrass, Über die analytische Darstellbarkeit sogenannter willkürlicher Functionen einer reellen Veränderlichen. Sitzungsberichte der Königlich Preußischen Akademie der Wissenschaften zu Berlin **2**, 633–639 (1885)
16. T. Yaying, B. Hazarika, S.A. Mohiuddine, A.A. Bakery, Norm inequalities involving upper bounds for certain matrix operators on Orlicz-Euler and Orlicz-Taylor double sequence spaces. J. Anal. **31**(4), 2817–2834 (2023)

# Fixed-Point Method: Stability Results of Functional Equation Deriving from Additive-Quadratic Mappings in 2-Banach Spaces

**Kandhasamy Tamilvanan, N. Revathi, Yahya Almalki, and N. Uthirasamy**

**Abstract** This work aims to investigate the Ulam stability of the quadratic-additive functional equation in 2-Banach spaces using direct method and fixed-point method. The study focuses on the stability of quadratic mapping and additive mapping. The results obtained have a practical applications, demonstrating that the stability of the quadratic-additive functional equation can be controlled through the sums and products of powers of norms.

**Keywords** Quadratic-additive functional equation · Ulam stability · Fixed point · 2-Banach spaces

**MSC** 39B52 · 39B72 · 39B82

---

K. Tamilvanan (✉)
Faculty of Science & Humanities, Department of Mathematics, R.M.K. Engineering College, Kavaraipettai, Tiruvallur 601206, Tamil Nadu, India
e-mail: tamiltamilk7@gmail.com

N. Revathi
Department of Computer Science and Applications, SRM Institute of Science and Technology, Chennai 600089, Tamil Nadu, India

Y. Almalki
Department of Mathematics, College of Science, King Khalid University, Abha 61413, Saudi Arabia
e-mail: yalmalki@kku.edu.sa

N. Uthirasamy
Department of Mathematics, K.S.Rangasamy College of Technology, Tiruchengode, Tamil Nadu, India
e-mail: uthirasamy@ksrct.ac.in



S. A. Mohiuddine et al. (eds.), *Positive Operators and Fixed-Point Theorems with Applications*, Industrial and Applied Mathematics,
https://doi.org/10.1007/978-981-96-2476-8_10

# 1 Introduction

The study of stability problems for functional equations is one of the important research fields in mathematics, which originated in issues related to applied mathematics. The first question concerning the stability of homomorphisms was given by Ulam [19] as follows.

Given a group $(G, *)$, a metric group $(G', \cdot)$ with the metric $d$, and a mapping $f$ from $G$ and $G'$, does $\delta > 0$ exist such that

$$d(f(x * y), f(x) \cdot f(y)) \leq \delta$$

for all $x, y \in G$. If such a mapping exists, then does a homomorphism $h : G \to G'$ exist such that

$$d(f(x), h(x)) \leq \epsilon$$

for all $x \in G$? Ulam defined such a problem in 1940 and solved it the following year for the Cauchy functional equation

$$f(x + y) = f(x) + f(y) \tag{1.1}$$

by the way of Hyers [7]. Subsequently, Aoki [2] and Rassias [11] extended Hyers' work for additive and linear mappings, respectively, by permitting the Cauchy difference to be unbounded. By substituting a general control function for the unbounded Cauchy difference, Găvruţă [4] proposed a generalization of the Rassias theorem. Numerous mathematicians have studied the Ulam–Hyers stability results and the expanded Ulam–Hyers stability results problem for various functional equations (cf. [3, 8, 12]).

In [1], the authors introduced a $\lambda$-quadratic functional equation and derived its general solution. They proceeded to analyze the Ulam–Hyers stability of this functional equation in a non-Archimedean Banach space, employing both direct method and fixed-point method. Additionally, they investigated the stability results in non-Archimedean random normed spaces.

In [9], Mohiuddine et al. introduced a novel class of generalized quartic functional equations and derived their general solution. They proceeded to investigate the stability results using the Hyers method in modular spaces for quartic functional equations. Particularly, their approach didn't rely on the Fatou property or the $\Delta_b$-condition, and they individually investigated stability without using both these conditions. Furthermore, they employed a fixed-point method incorporating the concept of the Fatou property in modular spaces to analyze stability for this functional equation. Additionally, a suitable counterexample was presented to illustrate the non-stability of a specific case.

In [13], the authors investigated the Hyers–Ulam stability results of the quadratic functional equation in both Banach spaces and non-Archimedean Banach spaces. They employed two different methods: direct method and fixed-point method, to investigate the stability of the functional equation. More functional equations may be found in [14–18].

This work aims to investigate the Ulam stability of the quadratic-additive functional equation:

$$\tau\left(\sum_{1\le a\le s} at_a\right)+\sum_{1\le a\le s}\tau\left(-at_a+\sum_{b=1;a\neq b}^{s} bt_b\right)$$
$$=(s-3)\sum_{1\le a<b\le s}\tau\left(at_a+bt_b\right) \tag{1.2}$$

$$-\left(s^2-5s+2\right)\sum_{1\le a\le s}a^2\left[\frac{\tau(t_a)+\tau(-t_a)}{2}\right]$$
$$-\left(s^2-5s+4\right)\sum_{1\le a\le s}a\left[\frac{\tau(t_a)-\tau(-t_a)}{2}\right], \tag{1.3}$$

where $s > 4$, in 2-Banach space using direct method and fixed-point method. The study focuses on the stability of additive mapping and quadratic mapping. The results obtain have practical applications, demonstrating that the stability of the mixed-type quadratic-additive functional equation can be controlled through the sums and products of powers of norms.

## 2 Preliminaries

In 1960, Gahler [5, 6] introduced the concept of linear 2-normed spaces.

**Definition 2.1** ([5, 6]) Consider a linear space $B$ over $\mathbb{R}$ with dim $B > 1$, and let $\|\cdot,\cdot\| : B^2 \to \mathbb{R}$ be a mapping that satisfies the following conditions:

(a) $\|x, y\| = 0$ holds if and only if $x$ and $y$ are linearly dependent,
(b) $\|x, y\| = \|y, x\|$,
(c) $\|\alpha x, y\| = |\alpha|\,\|x, y\|$,
(d) $\|x, y+z\| \le \|x, y\| + \|x, z\|$

for all $x, y, z \in B$ and all $\alpha \in \mathbb{R}$.

The function $\|\cdot,\cdot\|$ is called as a 2-norm on $B$, and the pair $(B, \|\cdot,\cdot\|)$ is said to be a linear 2-normed space. An illustrative example of a 2-normed space is $\mathbb{R}^2$ equipped with the 2-norm defined as $|x, y|$ equaling the area of the triangle with vertices at 0, $x$, and $y$.

As a consequence of property (d), it follows that

$$\|x + y, z\| \leq \|x, y\| + \|y, z\| \quad \text{and} \quad \big| \, \|x, z\| - \|y, z\| \, \big| \leq \|x - y, z\|.$$

Hence, for any fixed $y \in B$, the mapping $x \to \|x, y\|$ are continuous mappings from $B$ to $\mathbb{R}$.

**Definition 2.2** ([5, 6]) In a linear 2-normed space $B$, a sequence $\{x_j\}$ is called a Cauchy sequence if there exists two points $y$ and $z$ in $B$ such that $y$ and $z$ are linearly independent,

$$\lim_{i,j \to \infty} \|x_i - x_j, y\| = 0$$

and

$$\lim_{i,j \to \infty} \|x_i - x_j, z\| = 0.$$

**Definition 2.3** ([5, 6]) In a linear 2-normed space $B$, a sequence $\{x_j\}$ is said to be a convergent sequence if there exists an element $x \in B$ such that

$$\lim_{i,j \to \infty} \|x_j - x, y\| = 0$$

for all $y \in B$. If $\{x_j\}$ converges to $x$, then $x_j \to x$ is denoted as $j \to \infty$, and $x$ is said to be the limit point of $\{x_j\}$. In this case, we also write

$$\lim_{j \to \infty} x_j = x.$$

**Definition 2.4** ([5, 6]) Every Cauchy sequence convergent in a 2-Banach space is called a linear 2-normed space.

**Lemma 2.5** ([10]) *Let $(B, \|\cdot, \cdot\|)$ be a linear 2-normed space. For any $y \in B$, if $x$ is in $B$ and $\|x, y\| = 0$, then $x$ is equal to 0.*

**Lemma 2.6** ([10]) *In a linear 2-normed space $B$, with a convergent sequence $\{x_j\}$,*

$$\lim_{j \to \infty} \|x_j, y\| = \|\lim_{j \to \infty} x_j, y\|$$

*for all $y \in B$.*

The following fixed-point theorem plays an important role in the findings of the stability of the functional Eq. (1.2).

**Theorem 2.7** ([1]) *Let $(M, d)$ be a complete generalized metric space and a mapping $\Lambda : M \to M$ is strictly contractive with Lipschitz constant $L < 1$. Consider for an element $x \in M$, there exists an integer $i > 0$ satisfies $d(\Lambda^{i+1}x, \Lambda^i x) < \infty$. Then*

*(i) the sequence $\{\Lambda^i x\}_{i=1}^{\infty}$ converges to a fixed point $x^* \in M$ of $\Lambda$;*
*(ii) $x^*$ is the unique fixed point of $\Lambda$ in the set $S = \{y \in M : d(\Lambda^i x, y) < \infty\}$;*
*(iii) $d(y, x^*) \leq \frac{1}{1-L} d(y, \Lambda y)$ for all $y \in S$.*

For the notational handiness, let us define a mapping $\tau : B \to D$ as follows:

$$\begin{aligned} D\tau(t_1, t_2, \dots, t_s) = {} & \tau\left(\sum_{1\le a\le s} at_a\right) + \sum_{1\le a\le s} \tau\left(-at_a + \sum_{b=1;a\neq b}^{s} bt_b\right) \\ & -(s-3)\sum_{1\le a<b\le s} \tau\left(at_a + bt_b\right) \\ & +\left(s^2 - 5s + 2\right)\sum_{1\le a\le s} a^2\left[\frac{\tau(t_a)+\tau(-t_a)}{2}\right] \\ & +\left(s^2 - 5s + 4\right)\sum_{1\le a\le s} a\left[\frac{\tau(t_a)-\tau(-t_a)}{2}\right] \end{aligned}$$

for all $t_1, t_2, \dots, t_s \in B$.

## 3 Stability Results in 2-Banach Spaces

In this section, let us consider $B$ is a normed linear space and $D$ is a 2-Banach space.

### 3.1 Direct Method for Odd Case

**Theorem 3.1** *Let $\varphi : B \to [0, +\infty)$ be a mapping such that*

$$\lim_{i\to\infty} \frac{1}{2^i}\varphi\left(2^i t_1, 2^i t_2, \dots, 2^i t_s, l\right) = 0 \tag{3.1}$$

*for all $t_1, t_2, \dots, t_l \in B$ and all $l \in D$. Suppose that a mapping $\tau : B \to D$ with $\tau(0) = 0$ and such that*

$$\left\| D\tau(t_1, t_2, \dots, t_s), l \right\| \le \varphi(t_1, t_2, \dots, t_s, l) \tag{3.2}$$

*and*

$$\hat{\varphi}(t, l) =: \sum_{j=0}^{\infty} \frac{1}{2^j}\varphi(0, 2^j t, 0, \dots, 0, l) < \infty$$

*exists for all $t_1, t_2, \dots, t_s \in B$ and all $l \in D$. Then there exists an unique additive mapping $A_1 : B \to D$ satisfying*

$$\left\| \tau(t) - A_1(t), l \right\| \le \frac{1}{c}\hat{\varphi}(t, l), \tag{3.3}$$

*where* $c = (s^2 - 5s + 4)$, *for all* $t \in B$ *and all* $l \in D$.

***Proof*** Switching $(t_1, t_2, t_3, \ldots, t_s)$ by $(0, t, 0, \ldots, 0)$ in (3.2), we obtain

$$\left\| c\tau(2t) - 2c\tau(t), l \right\| \leq \varphi(0, t, 0, \ldots, 0, l), \tag{3.4}$$

for all $t \in B$ and all $l \in D$. Replacing $t$ by $2^n t$ in (3.4) and dividing by $2^{n+1}$, we have

$$\left\| \frac{\tau(2^{n+1}t)}{2^{(n+1)}} - \frac{\tau(2^n t)}{2^n}, l \right\| \leq \frac{1}{2^{n+1}c}\varphi(0, 2^n t, 0, \ldots, 0, l),$$

for all $t \in B$, $l \in D$ and all positive integers $n$. Hence,

$$\begin{aligned} \left\| \frac{\tau(2^{n+1}t)}{2^{n+1}} - \frac{\tau(2^m t)}{2^m}, l \right\| &\leq \sum_{j=m}^{n} \left\| \frac{\tau(2^{j+1}t)}{2^{j+1}} - \frac{\tau(2^j t)}{2^j}, l \right\| \\ &\leq \frac{1}{2c} \sum_{j=m}^{n} \frac{1}{2^j}\varphi(0, 2^j t, 0, \ldots, 0, l), \end{aligned} \tag{3.5}$$

for all $t \in B$, $l \in D$, and all integers $n \geq m > 0$. From inequalities (3.2) and (3.5), the sequence $\{\frac{\tau(2^i t)}{2^i}\}$ is a Cauchy in $D$ for all $t \in B$. Since $D$ is complete, the sequence $\{\frac{\tau(2^i t)}{2^i}\}$ converges in $D$ for all $t \in B$. Now, we can define a mapping $A_1 : B \to D$ by

$$A_1(t) := \lim_{i\to\infty} \frac{\tau(2^i t)}{2^i}, \tag{3.6}$$

for all $t \in B$. Thus,

$$\lim_{i\to\infty} \left\| \frac{\tau(2^i t)}{2^i} - A_1(t), l \right\| = 0$$

for all $t \in B$ and all $l \in D$. Setting $m = 0$ and taking the limit as $i \to \infty$ in inequality (3.5), we obtain (3.3). Next, we want to prove that the function $A_1$ is additive. From inequalities (3.1), (3.2), (3.6), and Lemma 2.6, that

$$\begin{aligned} \left\| D\tau(t_1, t_2, \ldots, t_s), l \right\| &= \lim_{i\to\infty} \left\| D\tau(2^i t_1, 2^i t_2, \ldots, 2^i t_s), l \right\| \\ &\leq \lim_{i\to\infty} \frac{1}{2^i}\varphi(2^i t_1, 2^i t_2, \ldots, 2^i t_s, l) = 0 \end{aligned}$$

for all $t_1, t_2, \ldots, t_s \in B$ and all $l \in D$. By Lemma 2.5,

$$DA_1(t_1, t_2, \ldots, t_s) = 0$$

for all $t_1, t_2, \ldots, t_s \in B$. Hence, the function $A_1 : B \to D$ is additive.

To verify that the function $A_1$ is unique, let us consider another additive mapping $A_1^{'} : B \to D$ satisfying the inequality (3.3). Then

$$\left\| A_1(t) - A_1^{'}(t), l \right\| = \lim_{i\to\infty} \frac{1}{2^i} \left\| A_1(2^i t) - \tau(2^i t) + \tau(2^i t) - A_1^{'}(2^i t), l \right\| \\ \le \frac{1}{c} \lim_{i\to\infty} \frac{\hat{\varphi}(2^i t, l)}{2^i} = 0$$

for all $t \in B$ and all $l \in D$. By Lemma 2.5, $A_1(t) - A_1^{'}(t) = 0$ for all $t \in B$. Hence, $A_1 = A_1^{'}$. □

**Remark 3.2** Analogous to Theorem 3.1, we can formulate a theorem where the sequence $A_1(t) := \lim_{i\to\infty} 2^i \tau\left(\frac{t}{2^i}\right)$ is defined, provided appropriate assumptions for $\varphi$.

**Corollary 3.3** *Let* $\tau : [0, \infty) \to [0, \infty)$ *be a mapping such that* $\tau(0) = 0$ *and*

*(i)* $\tau(vs) \le \tau(v)\tau(s)$.
*(ii)* $\tau(v) < v$ *for all* $v > 1$.

*If a mapping* $\tau : B \to D$ *with* $\tau(0) = 0$ *and*

$$\left\| D\tau(t_1, t_2, \ldots, t_s), l \right\| \le \sum_{i=1}^{s} \tau(\|t_i\|) + \tau(\|l\|)$$

*for all* $t_1, t_2, \ldots, t_s \in B$ *and all* $l \in D$, *then there exists a unique additive mapping* $A_1 : B \to D$ *satisfying*

$$\left\| \tau(t) - A_1(t), l \right\| \le \frac{1}{c} \left[ \frac{2\tau(\|t\|)}{2 - \tau(2)} + \tau(\|l\|) \right], \tag{3.7}$$

*where* $c = (s^2 - 5s + 4)$, *for all* $s \in B$ *and all* $l \in D$.

***Proof*** Setting

$$\varphi(t_1, t_2, \ldots, t_s, l) = \sum_{i=1}^{s} \tau(\|t_i\|) + \tau(\|l\|)$$

for all $t_1, t_2, \ldots, t_s \in B$ and all $l \in D$. From condition (i) that

$$\tau(2^i) \le (\tau(2))^i$$

and

$$\varphi(2^i t_1, 2^i t_2, \ldots, 2^i t_s, l) \le (\tau(2))^i \left( \sum_{i=1}^{s} \tau(\|t_i\|) \right) + \tau(\|l\|).$$

By using Theorem 3.1, we obtain (3.7). □

**Corollary 3.4** *Let $w$ be a positive real number such that $w < 1$, and let $S : [0, \infty) \times [0, \infty) \to [0, \infty)$ be a homogeneous mapping with degree $w$. Suppose a mapping $\tau : B \to D$ with $\tau(0) = 0$, and*

$$\left\| D\tau(t_1, t_2, \ldots, t_s), l \right\| \leq S\left(\|t_1\|, \|t_2\|, \ldots, \|t_s\|\right) + \|l\|$$

*for all $t_1, t_2, \ldots, t_s \in B$ and all $l \in D$, then there exists a unique additive mapping $A_1 : B \to D$ satisfying*

$$\left\| \tau(t) - A_1(t), l \right\| \leq \frac{S(0, \|t\|, 0, \ldots, 0) + \|l\|}{c(2 - w)}, \tag{3.8}$$

*where $c = (s^2 - 5s + 4)$, for all $t \in B$ and all $l \in D$.*

***Proof*** Let

$$\varphi(t_1, t_2, \ldots, t_s, l) = S\left(\|t_1\|, \|t_2\|, \ldots, \|t_s\|\right) + \|l\|$$

for all $t_1, t_2, \ldots, t_s \in B$ and all $l \in D$. By using Theorem 3.1, we obtain (3.8). □

**Corollary 3.5** *Let $w$ be a positive real number such that $w < 1$ and let $S : [0, \infty) \times [0, \infty) \to [0, \infty)$ a homogeneous function with degree $w$. If a mapping $\tau : B \to D$ with $\tau(0) = 0$ and*

$$\left\| D\tau(t_1, t_2, \ldots, t_s), l \right\| \leq S\left(\|t_1\|, \|t_2\|, \ldots, \|t_s\|\right) \|l\|$$

*for all $t_1, t_2, \ldots, t_s \in B$ and all $l \in D$, then there exists an unique additive mapping $A_1 : B \to D$ satisfying*

$$\left\| \tau(t) - A_1(t), l \right\| \leq \frac{S(0, \|t\|, 0, \ldots, 0)\|l\|}{2c(2 - 2^w)}, \tag{3.9}$$

*where $c = (s^2 - 5s + 4)$, for all $t \in B$ and all $l \in D$.*

***Proof*** Setting

$$\varphi(t_1, t_2, \ldots, t_s, l) = S\left(\|t_1\|, \|t_2\|, \ldots, \|t_s\|\right) \|l\|$$

for all $t_1, t_2, \ldots, t_s \in B$ and all $l \in D$. By using Theorem 3.1, we obtain (3.9). □

**Corollary 3.6** *Let $w$ be a positive real number such that $w < 1$. If a mapping $\tau : B \to D$ with $\tau(0) = 0$ and*

$$\left\| D\tau(t_1, t_2, \ldots, t_s), l \right\| \leq \sum_{i=1}^{s} \|t_i\|^w + \|l\|$$

*for all* $t_1, t_2, \ldots, t_s \in B$ *and all* $l \in D$*, then there exists an unique additive mapping* $A_1 : B \to D$ *satisfying*

$$\left\| \tau(t) - A_1(t), l \right\| \leq \frac{2\|t\|^w + \|l\|}{c(2 - w)},$$

*where* $c = (s^2 - 5s + 4)$*, for all* $t \in B$ *and all* $l \in D$.

## 3.2 Fixed-Point Method for Odd Case

**Theorem 3.7** *Let* $\tau : B \to D$ *be a mapping for which there exists a mapping* $\varphi : B^s \to [0, \infty)$ *and*

$$\lim_{w \to \infty} \frac{\varphi(\eta_g^w t_1, \eta_g^w t_2, \eta_g^w t_3, \ldots, \eta_g^w t_s, l)}{\eta_g^w} = 0, \tag{3.10}$$

*where* $\eta_g = \begin{cases} 2, & \text{if } g = 0, \\ \frac{1}{2} & \text{if } g = 1, \end{cases}$ *and such that it satisfies the inequality*

$$\|D\tau(t_1, t_2, t_3, \ldots, t_s), l\| \leq \varphi(t_1, t_2, t_3, \ldots, t_s, l) \tag{3.11}$$

*for all* $t_1, t_2, t_3, \ldots, t_s \in B$ *and all* $l \in D$*. If there exists a Lipschitz constant* $L = L(g)$ *such that*

$$t \to \lambda(t) = \frac{1}{c}\varphi\left(0, \frac{t}{2}, 0, \ldots, 0, l\right)$$

*has the property*

$$\frac{\lambda(\eta_g t)}{\eta_g} = L\lambda(t) \tag{3.12}$$

*for all* $t \in B$*. Then there exists an unique additive mapping* $A_1 : B \to D$ *satisfying the functional Eq. (1.2) and*

$$\|\tau(t) - A_1(t), l\| \leq \frac{L^{1-g}}{1 - L}\lambda(t), \tag{3.13}$$

*where* $c = (s^2 - 5s + 4)$*, for all* $t \in B$ *and all* $l \in D$.

***Proof*** Let us consider a set

$$\zeta = \{r/r : B \to D, r(0) = 0\}$$

along with initiating the generalized metric on $\zeta$,

$$d(r, n) = \inf\{w \in [0, \infty) : \|r(t) - n(t), l\| \leq w\lambda(t), \quad t \in B\}.$$

It is straightforward to observe that $(\zeta, d)$ is a complete space. Fix a mapping $\xi : \zeta \to \zeta$ as

$$\xi m(t) = \frac{1}{\eta_g} m(\eta_g t),$$

for all $t \in B$. For $m, r \in \zeta$ and $t \in B, l \in D$, we obtain

$$\begin{aligned} d(m, r) &= w \\ \Rightarrow \|m(t) - r(t), l\| &\leq w\lambda(t) \\ \Rightarrow \left\| \frac{m(\eta_g t)}{\eta_g} - \frac{r(\eta_g t)}{\eta_g}, l \right\| &\leq \frac{1}{\eta_g} w\lambda(\eta_g t) \\ \Rightarrow \|\xi m(t) - \xi r(t), l\| &\leq \frac{1}{\eta_g} w\lambda(\eta_g t) \\ \Rightarrow d(\xi m(t), \xi r(t)) &\leq wL. \end{aligned}$$

That is, $d(\xi m, \xi r) \leq Ld(m, r)$. By applying the fixed-point method, we establish that $\xi$ is a strictly contractive mapping on $\zeta$ with $L$. Replacing $(t_1, t_2, t_3, \ldots, t_s)$ by $(0, t, 0, \ldots, 0)$ in (3.2), which gives that

$$\|c\tau(2t) - 2c\tau(t), l\| \leq \varphi(0, t, 0, \ldots, 0, l), \tag{3.14}$$

for all $t \in B$ and all $l \in D$. From inequality (3.14), it follows that

$$\left\| \frac{\tau(2t)}{2} - \tau(t), l \right\| \leq \frac{\varphi(0, t, 0, \ldots, 0, l)}{2c}, \tag{3.15}$$

for all $t \in B$ and all $l \in D$. By using the inequality (3.12) with the case of $g = 0$, we obtain

$$\begin{aligned} \left\| \tau(t) - \frac{\tau(2t)}{2}, l \right\| &\leq \frac{1}{2}\lambda(t) \\ \Rightarrow \|\tau(t) - \tau m(t), l\| &\leq L\lambda(t), \end{aligned}$$

for all $t \in B$ and all $l \in D$. Thus, we obtain

$$d(\xi\tau(t), \tau(t)) \leq L = L^{1-g} < \infty, \tag{3.16}$$

for all $t \in B$. Replacing $t$ by $\frac{t}{2}$ in (3.15), we get

$$\left\| \tau(t) - 2\tau\left(\frac{t}{2}\right), l \right\| \leq \frac{1}{c}\varphi\left(0, \frac{t}{2}, 0, \ldots, 0, l\right),$$

for all $t \in B$. By employing the inequality (3.12) with $g = 1$, we get

$$\left\|2\tau\left(\frac{t}{2}\right) - \tau(t), l\right\| \le \lambda(t)$$
$$\Rightarrow \|\xi\tau(t) - \tau(t), l\| \le \lambda(t),$$

for all $t \in B$ and all $l \in D$. Consequently, we can conclude that

$$d(\tau(t), \xi\tau(t)) \le 1 = L^0 = L^{1-g}, \tag{3.17}$$

for all $t \in B$. By considering inequalities (3.16) and (3.17), we can deduce that

$$d(\tau(t), \xi\tau(t)) \le L^{1-g} < \infty,$$

for all $t \in B$. Next, using the fixed-point alternative Theorem 2.7, we can establish the existence of a fixed point $A_1$ of $\xi$ in $\zeta$, satisfying

$$A_1(t) = \lim_{w\to\infty} \frac{\tau(\eta_g^w t)}{\eta_g^w},$$

for all $t \in B$. To establish that $A_1 : B \to D$ satisfies the functional Eq. (1.2), we use a similar argument as in the proof of Theorem 3.1. Since $A_1$ is an unique fixed point of $\xi$ in the set $\Delta = \{\tau \in A_1/d(\tau, \zeta) < \infty\}$, $A_1$ is an unique solution such that

$$d(\tau, A_1) \le \frac{1}{1-L} d(\tau, \xi\tau)$$
$$\Rightarrow d(\tau, A_1) \le \frac{L^{1-g}}{1-L}$$
$$\Rightarrow \|\tau(t) - A_1(t), l\| \le \frac{L^{1-g}}{1-L}\lambda(t),$$

for all $t \in B$. Hence the proof. □

**Corollary 3.8** *Suppose $\epsilon_1$ and $\epsilon_2$ are positive real numbers. If a mapping $\tau : B \to D$ such that*

$$\|D\tau(t_1, t_2, t_3, \ldots, t_s), l\| \le \begin{cases} \epsilon_1 \ \|l\| \\ \epsilon_1\{\sum_{i=1}^{s} \|t_i\|^{\epsilon_2}\} + \|l\|, \\ \epsilon_1\{(\prod_{i=1}^{s} \|t_i\|^{\epsilon_2} + \sum_{i=1}^{s} \|t_i\|^{s\epsilon_2})\} \ \|l\|, \end{cases}$$

*for all $t_1, t_2, t_3, \ldots, t_s \in B$, then there exists an unique additive mapping $A_1 : B \to D$ satisfying*

$$\|\tau(t) - A_1(t), l\| \le \begin{cases} |\frac{\epsilon_1}{c}| \ \|l\|, \\ \frac{2\epsilon_1 \|t\|^{\epsilon_2} + \|l\|}{c|2-2^{\epsilon_2}|}, & \epsilon_2 \neq 1, \\ \frac{2\epsilon_1 \|t\|^{s\epsilon_2} \ \|l\|}{c|2-2^{s\epsilon_2}|}, & \epsilon_2 \neq \frac{1}{s}, \end{cases} \tag{3.18}$$

*where* $c = (s^2 - 5s + 4)$, *for all* $t \in B$.

***Proof*** Consider,

$$\varphi(t_1, t_2, t_3, \ldots, t_s, l) \leq \begin{cases} \epsilon_1 \ \|l\| \\ \epsilon_1\{\sum_{i=1}^{s} \|t_i\|^{\epsilon_2}\} + \|l\|, \\ \epsilon_1\{(\prod_{i=1}^{s} \|t_i\|^{\epsilon_2} + \sum_{i=1}^{s} \|t_i\|^{s\epsilon_2})\} \ \|l\|, \end{cases}$$

for all $t_1, t_2, t_3, \ldots, t_s \in B$. Now,

$$\frac{\varphi(\eta_g^w t_1, \eta_g^w t_2, \eta_g^w t_3, \ldots, \eta_g^w t_s, l)}{\eta_g^w} = \begin{cases} \frac{\epsilon_1}{\eta_g^w} \ \|l\|, \\ \frac{\epsilon_1}{\eta_g^w}\{\sum_{i=1}^{s} \|\eta_g t_i\|^{\epsilon_2}\} + \ \|l\|, \\ \frac{\epsilon_1}{\eta_g^w}\{(\prod_{i=1}^{s} \|\eta_g t_i\|^{\epsilon_2} + \sum_{i=1}^{s} \|\eta_g t_i\|^{s\epsilon_2})\} \ \|l\|, \end{cases}$$

$$= \begin{cases} \rightarrow 0 \quad as \quad w \rightarrow \infty, \\ \rightarrow 0 \quad as \quad w \rightarrow \infty, \\ \rightarrow 0 \quad as \quad w \rightarrow \infty, \end{cases}$$

i.e., the inequality (3.10) holds. Since we have

$$\lambda(t) = \frac{1}{c}\varphi\left(0, \frac{t}{2}, 0, \ldots, 0, l\right).$$

Also,

$$\lambda(t) = \frac{1}{c}\varphi\left(0, \frac{t}{2}, 0, \ldots, 0, l\right) = \begin{cases} \frac{\epsilon_1}{c} \ \|l\|, \\ \frac{2\epsilon_1\|t\|^{\epsilon_2} + \ \|l\|}{c2^{\epsilon_2}}, \\ \frac{2\epsilon_1\|t\|^{s\epsilon_2} \ \|l\|}{c2^{s\epsilon_2}} \end{cases}$$

and

$$\frac{1}{\eta_g}\lambda(\eta_g t) = \begin{cases} \frac{\epsilon_1 \ \|l\|}{c\eta_g}, \\ \frac{2}{\eta_g} \frac{\epsilon_1\|t\|^{\epsilon_2}\eta_g^{\epsilon_2} + \ \|l\|}{c2^{\epsilon_2}}, \\ \frac{2}{\eta_g} \frac{\epsilon_1\|t\|^{s\epsilon_2}\eta_g^{s\epsilon_2} \ \|l\|}{c2^{s\epsilon_2}}, \end{cases}$$

$$= \begin{cases} \eta_g^{-1}\lambda(t), \\ \eta_g^{\epsilon_2 - 1}\lambda(t), \\ \eta_g^{s\epsilon_2 - 1}\lambda(t), \end{cases}$$

for all $t \in B$. Now, by examining inequality (3.13), we can validate the following cases:

**Case 1:** $L = \frac{1}{2}$ if $g = 0$.

$$
\begin{aligned}
\|\tau(t) - A_1(t), l\| &\leq \frac{L^{1-g}}{1-L}\lambda(t) \\
&\leq \frac{\epsilon_1}{c(1-2^{-1})}\ \|l\| \\
\|\tau(t) - A_1(t), l\| &\leq \frac{\epsilon_1}{c}\ \|l\|.
\end{aligned}
$$

**Case 2:** $L = 2$ if $g = 1$.

$$
\begin{aligned}
\|\tau(t) - A_1(t), l\| &\leq \frac{L^{1-g}}{1-L}\lambda(t) \\
&\leq \frac{\epsilon_1}{c(1-2)}\ \|l\| \\
\|\tau(t) - A_1(t), l\| &\leq -\frac{\epsilon_1}{c}\ \|l\|.
\end{aligned}
$$

**Case 3:** $L = 2^{\epsilon_2 - 1}$ for $\epsilon_2 < 1$ if $g = 0$.

$$
\begin{aligned}
\|\tau(t) - A_1(t), l\| &\leq \frac{L^{1-g}}{1-L}\lambda(t) \\
&\leq \frac{2^{\epsilon_2-1}}{1-2^{\epsilon_2-1}}\ \frac{2\epsilon_1\|t\|_2^{\epsilon} + \ \|l\|}{c2^{\epsilon_2}} \\
\|\tau(t) - A_1(t), l\| &\leq \frac{2\epsilon_1\|t\|^{\epsilon_2} + \ \|l\|}{c(2-2^{\epsilon_2})}.
\end{aligned}
$$

**Case 4:** $L = 2^{1-\epsilon_2}$ for $\epsilon_2 > 1$ if $g = 1$.

$$
\begin{aligned}
\|\tau(t) - A_1(t), l\| &\leq \frac{L^{1-g}}{1-L}\lambda(t) \\
&\leq \frac{1}{1-2^{1-\epsilon_2}}\ \frac{2\epsilon_1\|t\|^{\epsilon_2} + \ \|l\|}{c2^{\epsilon_2}} \\
\|\tau(t) - A_1(t), l\| &\leq \frac{2\epsilon_1\|t\|^{\epsilon_2} + \ \|l\|}{c(2^{\epsilon_2}-2)}.
\end{aligned}
$$

**Case 5:** $L = 2^{s\epsilon_2 - 1}$ for $\epsilon_2 < \frac{1}{s}$ if $g = 0$.

$$
\begin{aligned}
\|\tau(t) - A_1(t), l\| &\leq \frac{L^{1-g}}{1-L}\lambda(t) \\
&\leq \frac{2^{s\epsilon_2-1}}{1-2^{s\epsilon_2-1}}\ \frac{2\epsilon_1\|t\|^{s\epsilon_2}\ \|l\|}{c2^{s\epsilon_2}} \\
\|\tau(t) - A_1(t), l\| &\leq \frac{2\epsilon_1\|t\|^{s\epsilon_2}\ \|l\|}{c(2-2^{s\epsilon_2})}.
\end{aligned}
$$

**Case 6:** $L = 2^{1-s\epsilon_2}$ for $\epsilon_2 > \frac{1}{s}$ if $g = 1$.

$$\begin{aligned}\|\tau(t) - A_1(t), l\| &\leq \frac{L^{1-g}}{1-L}\lambda(t) \\ &\leq \frac{1}{1-2^{1-s\epsilon_2}} \frac{2\epsilon_1 \|t\|^{s\epsilon_2}}{c2^{s\epsilon_2}\ \|l\|} \\ \|\tau(t) - A_1(t), l\| &\leq \frac{2\epsilon_1 \|t\|^{s\epsilon_2}\ \|l\|}{c(2^{s\epsilon_2}-2)}.\end{aligned}$$

Hence the proof is accomplished. □

## 3.3 Direct Method for Even Case

**Theorem 3.9** *Let a mapping* $\varphi : B \to [0, +\infty)$ *such that*

$$\lim_{i\to\infty} \frac{1}{2^{2i}}\varphi\left(2^i t_1, 2^i t_2, \ldots, 2^i t_s, l\right) = 0 \tag{3.19}$$

*for all* $t_1, t_2, \ldots, t_l \in B$ *and all* $l \in D$. *Suppose a mapping* $\tau : B \to D$ *with* $\tau(0) = 0$ *and satisfies*

$$\left\| D\tau(t_1, t_2, \ldots, t_s), l \right\| \leq \varphi(t_1, t_2, \ldots, t_s, l) \tag{3.20}$$

*and*

$$\hat{\varphi}(t, l) =: \sum_{j=0}^{\infty} \frac{1}{2^{2j}}\varphi(0, 2^j t, 0, \ldots, 0, l) < \infty$$

*exists for all* $t_1, t_2, \ldots, t_s \in B$ *and all* $l \in D$. *Then there exists an unique quadratic mapping* $Q_2 : B \to D$ *satisfying*

$$\left\| \tau(t) - Q_2(t), l \right\| \leq \frac{1}{c}\hat{\varphi}(t, l), \tag{3.21}$$

*where* $c = (s^2 - 5s + 2)$, *for all* $t \in B$ *and all* $l \in D$.

***Proof*** Replacing $(t_1, t_2, t_3, \ldots, t_s)$ by $(0, t, 0, \ldots, 0)$ in (3.20), we obtain

$$\left\| c\tau(2^2 t) - 2c\tau(t), l \right\| \leq \varphi(0, t, 0, \ldots, 0, l), \tag{3.22}$$

for all $t \in B$ and all $l \in D$. Replacing $t$ by $2^n t$ in (3.22) and dividing by $2^{2(n+1)}$, we get

$$\left\| \frac{\tau(2^{n+1}t)}{2^{2(n+1)}} - \frac{\tau(2^n t)}{2^{2n}}, l \right\| \leq \frac{1}{2^{2(n+1)}c}\varphi(0, 2^n t, 0, \ldots, 0, l),$$

for all $t \in B, l \in D$ and all positive integers $n$. Hence,

$$\left\| \frac{\tau(2^{n+1}t)}{2^{2(n+1)}} - \frac{\tau(2^m t)}{2^{2m}}, l \right\| \le \sum_{j=m}^{n} \left\| \frac{\tau(2^{j-1}t)}{2^{2(j+1)}} - \frac{\tau(2^j t)}{2^{2j}}, l \right\|$$
$$\le \frac{1}{2c} \sum_{j=m}^{n} \frac{1}{2^{2j}} \varphi(0, 2^j t, 0, \ldots, 0, l), \qquad (3.23)$$

for all $t \in B, l \in D$, and all integers $n \ge m > 0$. From inequalities (3.20) and (3.23), the sequence $\{\frac{\tau(2^i t)}{2^{2i}}\}$ is a Cauchy in $D$ for all $t \in B$. Since $D$ is complete, the sequence $\{\frac{\tau(2^i t)}{2^{2i}}\}$ converges in $D$ for all $t \in B$. Now, we can define a mapping $Q_2 : B \to D$ by

$$Q_2(t) := \lim_{i \to \infty} \frac{\tau(2^i t)}{2^{2i}}, \qquad (3.24)$$

for all $t \in B$. Thus,

$$\lim_{i \to \infty} \left\| \frac{\tau(2^i t)}{2^{2i}} - Q_2(t), l \right\| = 0$$

for all $t \in B$ and all $l \in D$. Setting $m = 0$ and taking the limit as $i \to \infty$ in inequality (3.23), we obtain (3.21). Next, we want to prove that the function $Q_2$ is quadratic. From inequalities (3.19), (3.20), (3.24), and Lemma 2.6, that

$$\left\| D\tau(t_1, t_2, \ldots, t_s), l \right\| = \lim_{i \to \infty} \left\| D\tau(2^i t_1, 2^i t_2, \ldots, 2^i t_s), l \right\|$$
$$\le \lim_{i \to \infty} \frac{1}{2^{2i}} \varphi(2^i t_1, 2^i t_2, \ldots, 2^i t_s, l) = 0$$

for all $t_1, t_2, \ldots, t_s \in B$ and all $l \in D$. By Lemma 2.5,

$$DQ_2(t_1, t_2, \ldots, t_s) = 0$$

for all $t_1, t_2, \ldots, t_s \in B$. Hence, the function $Q_2 : B \to D$ is quadratic.

To verify that the function $Q_2$ is unique, let us consider another quadratic mapping $Q_2' : B \to D$ satisfying the inequality (3.21). Then

$$\left\| Q_2(t) - Q_2'(t), l \right\| = \lim_{i \to \infty} \frac{1}{2^{2i}} \left\| Q_2(2^i t) - \tau(2^i t) + \tau(2^i t) - Q_2'(2^i t), l \right\|$$
$$\le \frac{1}{c} \lim_{i \to \infty} \frac{1}{2^{2i}} \hat{\varphi}(2^i t, l) = 0$$

for all $t \in B$ and all $l \in D$. By Lemma 2.5, we obtain

$$Q_2(t) - Q_2'(t) = 0$$

for all $t \in B$. Hence, $Q_2 = Q_2'$. □

**Remark 3.10** Analogous to Theorem 3.9, we can formulate a theorem where the sequence $Q_2(t) := \lim_{i\to\infty} 2^{2i}\tau\left(\frac{t}{2^i}\right)$ is defined, provided appropriate assumptions for $\varphi$.

**Corollary 3.11** *Let* $\tau : [0,\infty) \to [0,\infty)$ *be a mapping such that* $\tau(0) = 0$ *and*

*(i)* $\tau(vs) \le \tau(v)\tau(s)$.
*(ii)* $\tau(v) < v^2$ *for all* $v > 1$.

*If a mapping* $\tau : B \to D$ *with* $\tau(0) = 0$ *and*

$$\left\| D\tau(t_1, t_2, \ldots, t_s), l \right\| \le \sum_{i=1}^{s} \tau(\|t_i\|) + \tau(\|l\|)$$

*for all* $t_1, t_2, \ldots, t_s \in B$ *and all* $l \in D$*, then there exists an unique quadratic mapping* $Q_2 : B \to D$ *satisfying*

$$\left\| \tau(t) - Q_2(t), l \right\| \le \frac{1}{c}\left[ \frac{2\tau(\|t\|)}{2^2 - \tau(2)} + \tau(\|l\|) \right], \tag{3.25}$$

*where* $c = (s^2 - 5s + 2)$*, for all* $s \in B$ *and all* $l \in D$.

***Proof*** Setting

$$\varphi(t_1, t_2, \ldots, t_s, l) = \sum_{i=1}^{s} \tau(\|t_i\|) + \tau(\|l\|)$$

for all $t_1, t_2, \ldots, t_s \in B$ and all $l \in D$. From condition (i) that

$$\tau(2^i) \le (\tau(2))^i$$

and

$$\varphi(2^i t_1, 2^i t_2, \ldots, 2^i t_s, l) \le (\tau(2))^i \left( \sum_{i=1}^{s} \tau(\|t_i\|) \right) + \tau(\|l\|).$$

By using Theorem 3.9, we obtain (3.25). □

**Corollary 3.12** *Let* $w$ *be a positive real number such that* $w < 2$*, and let* $S : [0,\infty) \times [0,\infty) \to [0,\infty)$ *be a homogeneous mapping with degree* $w$*. If* $\tau : B \to D$ *be a mapping with* $\tau(0) = 0$ *and*

$$\left\| D\tau(t_1, t_2, \ldots, t_s), l \right\| \le S\left(\|t_1\|, \|t_2\|, \ldots, \|t_s\|\right) + \|l\|$$

*for all* $t_1, t_2, \ldots, t_s \in B$ *and all* $l \in D$*, then there exists a unique quadratic mapping* $Q_2 : B \to D$ *satisfying*

$$\left\| \tau(t) - Q_2(t), l \right\| \leq \frac{S(0, \|t\|, 0, \ldots, 0) + \|l\|}{c(2^2 - w^2)}, \tag{3.26}$$

*where* $c = (s^2 - 5s + 2)$, *for all* $t \in B$ *and all* $l \in D$.

***Proof*** Let

$$\varphi(t_1, t_2, \ldots, t_s, l) = S(\|t_1\|, \|t_2\|, \ldots, \|t_s\|) + \|l\|$$

for all $t_1, t_2, \ldots, t_s \in B$ and all $l \in D$. By using Theorem 3.9, we obtain (3.26). □

**Corollary 3.13** *Let* $w$ *be a positive real number such that* $w < 2$ *and let* $S : [0, \infty) \times [0, \infty) \to [0, \infty)$ *be a homogeneous function with degree* $w$. *If* $\tau : B \to D$ *be a mapping with* $\tau(0) = 0$ *and*

$$\left\| D\tau(t_1, t_2, \ldots, t_s), l \right\| \leq S(\|t_1\|, \|t_2\|, \ldots, \|t_s\|) \|l\|$$

*for all* $t_1, t_2, \ldots, t_s \in B$ *and all* $l \in D$, *then there exists a unique quadratic mapping* $Q_2 : B \to D$ *satisfying*

$$\left\| \tau(t) - Q_2(t), l \right\| \leq \frac{S(0, \|t\|, 0, \ldots, 0)\|l\|}{2c(2^2 - 2^w)}, \tag{3.27}$$

*where* $c = (s^2 - 5s + 2)$, *for all* $t \in B$ *and all* $l \in D$.

***Proof*** Setting

$$\varphi(t_1, t_2, \ldots, t_s, l) = S(\|t_1\|, \|t_2\|, \ldots, \|t_s\|) \|l\|$$

for all $t_1, t_2, \ldots, t_s \in B$ and all $l \in D$. By using Theorem 3.9, we obtain (3.27). □

**Corollary 3.14** *Let* $w$ *be a positive real number such that* $w < 2$. *If a mapping* $\tau : B \to D$ *with* $\tau(0) = 0$ *and*

$$\left\| D\tau(t_1, t_2, \ldots, t_s), l \right\| \leq \sum_{i=1}^{s} \|t_i\|^w + \|l\|$$

*for all* $t_1, t_2, \ldots, t_s \in B$ *and all* $l \in D$, *then there exists a unique quadratic mapping* $Q_2 : B \to D$ *satisfying*

$$\left\| \tau(t) - Q_2(t), l \right\| \leq \frac{2\|t\|^w + \|l\|}{c(2^2 - w^2)},$$

*where* $c = (s^2 - 5s + 2)$, *for all* $t \in B$ *and all* $l \in D$.

## 3.4 Fixed-Point Method for Even Case

**Theorem 3.15** *Let $\tau : B \to D$ be a mapping for which there exists a mapping $\varphi : B^s \to [0, \infty)$ and*

$$\lim_{w\to\infty} \frac{\varphi(\eta_g^w t_1, \eta_g^w t_2, \eta_g^w t_3, \dots, \eta_g^w t_s, l)}{\eta_g^{2w}} = 0, \tag{3.28}$$

*where* $\eta_g = \begin{cases} 2, & \text{if } g = 0, \\ \frac{1}{2} & \text{if } g = 1, \end{cases}$ *and such that it satisfies the inequality:*

$$\|D\tau(t_1, t_2, t_3, \dots, t_s), l\| \le \varphi(t_1, t_2, t_3, \dots, t_s, l)$$

*for all $t_1, t_2, t_3, \dots, t_s \in B$ and all $l \in D$. If there exists a Lipschitz constant $L = L(g)$ such that*

$$t \to \lambda(t) = \frac{1}{c}\varphi\left(0, \frac{t}{2}, 0, \dots, 0, l\right)$$

*has the property*

$$\frac{\lambda(\eta_g t)}{\eta_g^2} = L\lambda(t) \tag{3.29}$$

*for all $t \in B$. Then there exists a unique quadratic mapping $Q_2 : B \to D$ satisfying the functional Eq. (1.2) and*

$$\|\tau(t) - Q_2(t), l\| \le \frac{L^{1-g}}{1-L}\lambda(t), \tag{3.30}$$

*where $c = (s^2 - 5s + 2)$, for all $t \in B$ and all $l \in D$.*

***Proof*** Let us consider a set

$$\zeta = \{r/r : B \to D, r(0) = 0\}$$

along with initiating the generalized metric on $\zeta$,

$$d(r, n) = \inf\{w \in [0, \infty) : \|r(t) - n(t), l\| \le w\lambda(t), t \in B\}.$$

It is straightforward to observe that $(\zeta, d)$ is a complete space. Fix a mapping $\xi : \zeta \to \zeta$ as

$$\xi m(t) = \frac{m(\eta_g t)}{\eta_g^2},$$

for all $t \in B$. For $m, r \in \zeta$ and $t \in B, l \in D$, we obtain

$$
\begin{aligned}
& d(m, r) = w \\
\Rightarrow\ & \|m(t) - r(t), l\| \le w\lambda(t) \\
\Rightarrow\ & \left\| \frac{m(\eta_g t)}{\eta_g^2} - \frac{r(\eta_g t)}{\eta_g^2}, l \right\| \le \frac{1}{\eta_g^2} w\lambda(\eta_g t) \\
\Rightarrow\ & \|\xi m(t) - \xi r(t), l\| \le \frac{1}{\eta_g^2} w\lambda(\eta_g t) \\
\Rightarrow\ & d(\xi m(t), \xi r(t)) \le wL.
\end{aligned}
$$

That is, $d(\xi m, \xi r) \le L d(m, r)$. By applying the fixed-point method, we establish that $\xi$ is a strictly contractive mapping on $\zeta$ with $L$. Replacing $(t_1, t_2, t_3, \ldots, t_s)$ by $(0, t, 0, \ldots, 0)$ in inequality (3.20), which gives that

$$
\|c\tau(2t) - 2^2 c\tau(t), l\| \le \varphi(0, t, 0, \ldots, 0, l), \tag{3.31}
$$

for all $t \in B$ and all $l \in D$. From inequality (3.31), it follows that

$$
\left\| \frac{\tau(2t)}{2^2} - \tau(t), l \right\| \le \frac{\varphi(0, t, 0, \ldots, 0, l)}{2^2}, \tag{3.32}
$$

for all $t \in B$ and all $l \in D$. By using the inequality (3.29) with the case of $g = 0$, we obtain

$$
\begin{aligned}
& \left\| \tau(t) - \frac{\tau(2t)}{2^2}, l \right\| \le \frac{1}{2^2} \lambda(t) \\
\Rightarrow\ & \|\tau(t) - \tau m(t), l\| \le L\lambda(t),
\end{aligned}
$$

for all $t \in B$ and all $l \in D$. As a result, we obtain

$$
d(\xi\tau(t), \tau(t)) \le L = L^{1-g} < \infty, \tag{3.33}
$$

for all $t \in B$. Replacing $t$ by $\frac{t}{2}$ in (3.32), we obtain

$$
\left\| \tau(t) - 2^2 \tau\left(\frac{t}{2}\right), l \right\| \le \varphi\left(0, \frac{t}{2}, 0, \ldots, 0, l\right),
$$

for all $t \in B$ and all $l \in D$. By employing the inequality (3.29) with $g = 1$, we get

$$
\begin{aligned}
& \left\| 2^2 \tau\left(\frac{t}{2}\right) - \tau(t), l \right\| \le \lambda(t) \\
\Rightarrow\ & \|\xi\tau(t) - \tau(t), l\| \le \lambda(t),
\end{aligned}
$$

for all $t \in B$ and all $l \in D$. Consequently, we can conclude that

$$d(\tau(t), \xi\tau(t)) \leq 1 = L^0 = L^{1-g}, \tag{3.34}$$

for all $t \in B$. By considering inequalities (3.33) and (3.34), we can deduce that

$$d(\tau(t), \xi\tau(t)) \leq L^{1-g} < \infty,$$

for all $t \in B$. Next, using the fixed-point alternative Theorem 2.7, we can establish the existence of a fixed point $Q_2$ of $\xi$ in $\zeta$, satisfying

$$Q_2(t) = \lim_{w\to\infty} \frac{\tau(\eta_g^w t)}{\eta_g^{2w}},$$

for all $t \in B$. To establish that $Q_2 : B \to D$ satisfies the functional Eq. (1.2), we use a similar argument as in the proof of Theorem 3.9. Since $Q_2$ is an unique fixed point of $\xi$ in the set $\Delta = \{\tau \in Q_2/d(\tau, \zeta) < \infty\}$, $Q_2$ is the unique solution such that

$$\begin{aligned} d(\tau, Q_2) &\leq \frac{1}{1-L} d(\tau, \xi\tau) \\ \Rightarrow d(\tau, Q_2) &\leq \frac{L^{1-g}}{1-L} \\ \Rightarrow \|\tau(t) - Q_2(t), l\| &\leq \frac{L^{1-g}}{1-L}\lambda(t), \end{aligned}$$

for all $t \in B$. Hence the proof. □

**Corollary 3.16** *Suppose $\epsilon_1$ and $\epsilon_2$ are the positive real numbers. If a mapping $\tau : B \to D$ such that*

$$\|D\tau(t_1, t_2, t_3, \ldots, t_s), l\| \leq \begin{cases} \epsilon_1 \ \|l\|, \\ \epsilon_1\{\sum_{i=1}^{s} \|t_i\|^{\epsilon_2} + \ \|l\|\}, \\ \epsilon_1\{\left(\prod_{i=1}^{s} \|t_i\|^{\epsilon_2} + \sum_{i=1}^{s} \|t_i\|^{s\epsilon_2}\right) \ \|l\|\}, \end{cases}$$

*for all $t_1, t_2, t_3, \ldots, t_s \in B$ and all $l \in D$, then there exists an unique quadratic mapping $Q_2 : B \to D$ satisfying*

$$\|\tau(t) - Q_2(t), l\| \leq \begin{cases} |\frac{\epsilon_1}{3c}| \ \|l\| \\ \frac{2\epsilon_1 \|t\|^{\epsilon_2} + \ \|l\|}{c|2^2 - 2^{\epsilon_2}|}, & \epsilon_2 \neq 2, \\ \frac{2\epsilon_1 \|t\|^{s\epsilon_2} \ \|l\|}{c|2^2 - 2^{s\epsilon_2}|}, & \epsilon_2 \neq \frac{2}{s}, \end{cases} \tag{3.35}$$

*where $c = (s^2 - 5s + 2)$, for all $t \in B$ and all $l \in D$.*

***Proof*** Setting

$$\varphi(t_1, t_2, t_3, \ldots, t_s, l) \leq \begin{cases} \epsilon_1 \ \|l\|, \\ \epsilon_1\{\sum_{i=1}^{s} \|t_i\|^{\epsilon_2} + \ \|l\|\}, \\ \epsilon_1\{\left(\prod_{i=1}^{s} \|t_i\|^{\epsilon_2} + \sum_{i=1}^{s} \|t_i\|^{s\epsilon_2}\right) \ \|l\|\}, \end{cases}$$

for all $t_1, t_2, t_3, \ldots, t_s \in B$ and all $l \in D$. Now,

$$\frac{\varphi(\eta_g^w t_1, \eta_g^w t_2, \eta_g^w t_3, \ldots, \eta_g^w t_s, l)}{\eta_g^{2w}} = \begin{cases} \frac{\epsilon_1 \ \|l\|}{\eta_g^{2w}}, \\ \frac{\epsilon_1}{\eta_g^{2w}}\{\sum_{i=1}^{s} \|\eta_g t_i\|^{\epsilon_2} + \ \|l\|\}, \\ \frac{\epsilon_1}{\eta_g^{2w}}\{\left(\prod_{i=1}^{s} \|\eta_g t_i\|^{\epsilon_2} + \sum_{i=1}^{s} \|\eta_g t_i\|^{s\epsilon_2}\right) \ \|l\|\}, \end{cases}$$

$$= \begin{cases} \to 0 \quad as \quad w \to \infty, \\ \to 0 \quad as \quad w \to \infty, \\ \to 0 \quad as \quad w \to \infty, \end{cases}$$

i.e., the inequality (3.28) holds. Since we have

$$\lambda(t) = \frac{1}{c}\varphi\left(0, \frac{t}{2}, 0, \ldots, 0, l\right).$$

Also,

$$\lambda(t) = \frac{1}{c}\varphi\left(0, \frac{t}{2}, 0, \ldots, 0, l\right) = \begin{cases} \frac{\epsilon_1 \ \|l\|}{c}, \\ \frac{2\epsilon_1 \|t\|^{\epsilon_2} + \ \|l\|}{2^{\epsilon_2} c}, \\ \frac{2\epsilon_1 \|t\|^{s\epsilon_2} \ \|l\|}{2^{s\epsilon_2} c}, \end{cases}$$

and

$$\frac{1}{\eta_g^2}\lambda(\eta_g t) = \begin{cases} \frac{\epsilon_1 \ \|l\|}{c\eta_g^2}, \\ \frac{2}{\eta_g^2} \frac{\epsilon_1 \|t\|^{\epsilon_2} \eta_g^{\epsilon_2} + \ \|l\|}{2^{\epsilon_2} c}, \\ \frac{2}{\eta_g^2} \frac{\epsilon_1 \|t\|^{s\epsilon_2} \eta_g^{s\epsilon_2} \ \|l\|}{2^{s\epsilon_2} c}, \end{cases}$$

$$= \begin{cases} \eta_g^{-2}\lambda(t), \\ \eta_g^{\epsilon_2 - 2}\lambda(t), \\ \eta_g^{s\epsilon_2 - 2}\lambda(t), \end{cases}$$

for all $t \in B$. Now, by examining inequality (3.30), we can validate the following cases:

**Case 1**: $L = \frac{1}{2^2}$ if $g = 0$.

$$\begin{aligned}\|\tau(t) - Q_2(t), l\| &\leq \frac{L^{1-g}}{1-L}\lambda(t)\\ &\leq \frac{\epsilon_1 \ \|l\|}{c(1-2^{-2})}\\ \|\tau(t) - Q_2(t), l\| &\leq \frac{\epsilon_1 \ \|l\|}{3c}.\end{aligned}$$

**Case 2**: $L = 2^2$ if $g = 1$.

$$\begin{aligned}\|\tau(t) - Q_2(t), l\| &\leq \frac{L^{1-g}}{1-L}\lambda(t)\\ &\leq \frac{\epsilon_1 \ \|l\|}{c(1-2^{2})}\\ \|\tau(t) - Q_2(t), l\| &\leq -\frac{\epsilon_1 \ \|l\|}{3c}.\end{aligned}$$

**Case 3**: $L = 2^{\epsilon_2 - 2}$ for $\epsilon_2 < 2$ if $g = 0$.

$$\begin{aligned}\|\tau(t) - Q_2(t), l\| &\leq \frac{L^{1-g}}{1-L}\lambda(t)\\ &\leq \frac{2^{\epsilon_2-2}}{1-2^{\epsilon_2-2}} \frac{2\epsilon_1 \|t\|^{\epsilon_2} + \|l\|}{2^{\epsilon_2} c}\\ \|\tau(t) - Q_2(t), l\| &\leq \frac{2\epsilon_1 \|t\|^{\epsilon_2} + \|l\|}{(2^2 - 2^{\epsilon_2})c}.\end{aligned}$$

**Case 4**: $L = 2^{2-\epsilon_2}$ for $\epsilon_2 > 2$ if $g = 1$.

$$\begin{aligned}\|\tau(t) - Q_2(t), l\| &\leq \frac{L^{1-g}}{1-L}\lambda(t)\\ &\leq \frac{1}{1-2^{2-\epsilon_2}} \frac{2\epsilon_1 \|t\|^{\epsilon_2} + \|l\|}{2^{\epsilon_2} c}\\ \|\tau(t) - Q_2(t), l\| &\leq \frac{2\epsilon_1 \|t\|^{\epsilon_2} + \|l\|}{c(2^{\epsilon_2} - 2^2)}.\end{aligned}$$

**Case 5**: $L = 2^{s\epsilon_2 - 2}$ for $\epsilon_2 < \frac{2}{s}$ if $g = 0$.

$$\begin{aligned}\|\tau(t) - Q_2(t), l\| &\leq \frac{L^{1-g}}{1-L}\lambda(t)\\ &\leq \frac{2^{s\epsilon_2-2}}{1-2^{s\epsilon_2-2}} \frac{2\epsilon_1 \|t\|^{s\epsilon_2} \ \|l\|}{2^{s\epsilon_2} c}\end{aligned}$$

$$\|\tau(t) - Q_2(t), l\| \leq \frac{2\epsilon_1 \|t\|^{s\epsilon_2} \|l\|}{c(2^2 - 2^{s\epsilon_2})}.$$

**Case 6**: $L = 2^{2-s\epsilon_2}$ for $\epsilon_2 > \frac{2}{s}$ if $g = 1$.

$$\begin{aligned}\|\tau(t) - Q_2(t), l\| &\leq \frac{L^{1-g}}{1-L}\lambda(t)\\ &\leq \frac{1}{1-2^{2-s\epsilon_2}} \frac{2\epsilon_1 \|t\|^{s\epsilon_2} \|l\|}{2^{s\epsilon_2} c}\\ \|\tau(t) - Q_2(t), l\| &\leq \frac{2\epsilon_1 \|t\|^{s\epsilon_2} \|l\|}{c(2^{s\epsilon_2} - 2^2)}.\end{aligned}$$

Hence the proof is accomplished. □

## 4 Conclusion

In this work, we have investigated Ulam stability of a mixed-type quadratic-additive functional Equation (1.2) in 2-Banach spaces using direct method and fixed-point method. The study focused on the stability of additive mapping and quadratic mapping. The results obtained have practical applications, demonstrating that the stability of the quadratic-additive functional equation can be controlled through the sums and products of powers of norms.

## References


1. L.V. An, K. Tamilvanan, R. Udhayakumar, M.J. Kabeto, L.V. Ngoc, Ulam-Hyers stability results of $\lambda$-quadratic functional equation with three variables in non-Archimedean Banach space and non-Archimedean random normed space. J. Funct. Spaces **2022**, Art. ID 6795978, 10 pages (2022)
2. T. Aoki, On the stability of the linear transformation in Banach spaces. J. Math. Soc. Jpn. **2**, 64–66 (1950)
3. S. Czerwik, On the stability of the quadratic mapping in normed spaces. Abh. Math. Sem. Univ. Hamburg **62**, 59–64 (1992)
4. P. Găvruţă, A generalization of the Hyers-Ulam-Rassias stability of approximately additive mappings. J. Math. Anal. Appl. **184**(3), 431–436 (1994)
5. S. Gähler, 2-metrische Räume und ihre topologische Struktur. Math. Nachr. **26**, 115–148 (1963)
6. S. Gähler, Lineare 2-normierte Räume. Math. Nachr. **28**, 1–43 (1964)
7. D.H. Hyers, On the stability of the linear functional equation. Proc. Nat. Acad. Sci. USA **27**, 222–224 (1941)
8. S.-M. Jung, P.K. Sahoo, Hyers-Ulam stability of the quadratic equation of Pexider type. J. Korean Math. Soc. **38**(3), 645–656 (2001)
9. S.A. Mohiuddine, K. Tamilvanan, M. Mursaleen, T. Alotaibi, Stability of quartic functional equation in modular spaces via Hyers and fixed-point methods. Mathematics **10**(11), 22 pages (2022)

10. W.-G. Park, Approximate additive mappings in 2-Banach spaces and related topics. J. Math. Anal. Appl. **376**(1), 193–202 (2011)
11. T.M. Rassias, On the stability of the linear mapping in Banach spaces. Proc. Am. Math. Soc. **72**(2), 297–300 (1978)
12. F. Skof, Local properties and approximation of operators. Rend. Sem. Mat. Fis. Milano **53**(1983), 113–129 (1986)
13. K. Tamilvanan, A.M. Alanazi, M.G. Alshehri, J. Kafle, Hyers-Ulam stability of quadratic functional equation based on fixed point technique in Banach spaces and non-Archimedean Banach spaces. Mathematics **9**(20), 15 pages (2021)
14. K. Tamilvanan, A.H. Alkhaldi, R.P. Agarwal, A.M. Alanazi, Fixed point approach: Ulam stability results of functional equation in non-Archimedean fuzzy $\varphi$-2-normed spaces and non-Archimedean Banach spaces. Mathematics **11**(2), 24 pages (2023)
15. K. Tamilvanan, A.H. Alkhaldi, J. Jakhar, R. Chugh, J. Jakhar, J.M. Rassias, Ulam stability results of functional equations in modular spaces and 2-Banach spaces. Mathematics, **11**(2), 24 pages (2023)
16. K. Tamilvanan, Y. Almalki, S. A. Mohiuddine and R. P. Agarwal, Stability results of quadratic-additive functional equation based on Hyers technique in matrix paranormed spaces. Mathematics **10**(11), 17 pages (2022)
17. K. Tamilvanan, T. Alqahtani Rubayyi, S.A. Mohiuddine, Stability results of mixed type quadratic-additive functional equation in $\beta$-Banach modules by using fixed-point technique. Mathematics **10**(3), 21 pages (2022)
18. N. Uthirasamy, K. Tamilvanan, M.J. Kabeto, Ulam stability and non-stability of additive functional equation in IFN-spaces and 2-Banach spaces by different methods. J. Funct. Spaces **2022**, Article ID. 8028634, 14 pages (2022)
19. Ulam, S.M. A collection of mathematical problems, in *Interscience Tracts in Pure and Applied Mathematics*, no. 8 (Interscience Publishers, New York, 1960)

GPSR Compliance

*The European Union's (EU) General Product Safety Regulation (GPSR) is a set of rules that requires consumer products to be safe and our obligations to ensure this.*

*If you have any concerns about our products, you can contact us on ProductSafety@springernature.com*

In case Publisher is established outside the EU, the EU authorized representative is:

Springer Nature Customer Service Center GmbH
Europaplatz 3
69115 Heidelberg, Germany

**Batch number: 08998861**

Printed by Printforce, the Netherlands